A-Z
MOLECULAR PHYSICS

A-Z MOLECULAR PHYSICS

Prof. G.K. Bose

CENTRUM PRESS
NEW DELHI-110002 (INDIA)

CENTRUM PRESS
H.O.: 4360/4, Ansari Road, Daryaganj,
New Delhi-110 002 (India)
Ph.: 23278000, 23261597

B.O.: No. 1015, Ist Main Road, BSK IIIrd Stage
IIIrd Phase, IIIrd Block,
Bangalore - 560 085 (India)
Tel.: 080-41723429
Visit us at: www.centrumpress.com

A-Z Molecular Physics

First Edition, 2009

ISBN 978-93-80106-29-8

PRINTED IN INDIA

Printed at Salasar Imaging Systems, Delhi-110035 (India)

Contents

Preface

The introduction to this book surveys the general aspects of atomic cluster science and outlines some of its important new challenges. It contains an important definition of a cluster, as a new physical system possessing its own specific properties and features. This definition is important to establish that atomic cluster science is a new field of modern physics in its own right. It is highly multidisciplinary and has numerous links with traditional branches of physics and chemistry. The book concentrates on the theoretical aspects of molecular physics, such as the vibration, rotation, electronic states, potential curves, and spectra of molecules.

The different methods of approximation for the calculation of electronic wave functions and their energy are also covered. The introduction of basics terms used in group theory and their meaning in molecular physics enables an elegant description of polyatomic molecules and their symmetries. Molecular spectra and the dynamic processes involved in their excited states are given its own chapter.

The theoretical part then concludes with a discussion of the field of Van der Waals molecules and clusters. This book is devoted entirely to experimental techniques, such as laser, Fourier, NMR, and ESR spectroscopies, used in the fields of physics, chemistry, biology, and material science.

Author

Chapter 1

Atoms and Atomic Spectra

EARLY GREEK IDEAS

The first "atomic theorists" we have any record of were two fifth-century BC Greeks, Leucippus of Miletus and Democritus of Abdera. Their theories were naturally more philosophical than experimental in origin. The basic idea was that if you could look at matter on smaller and smaller scales ultimately you would see individual *atoms* - objects that could not be divided further. Everything was made up of these atoms, which moved around in a void (a vacuum). The different physical propertiescolor, taste, and so onof materials came about because atoms in them had different shapes and/or arrangements and orientations with respect to each other.

This was all pure conjecture, but the physical pictures they described sometimes seem uncannily accurate. For example, here is a quote from Lucretius, a contemporary of Julius Caesar, on the ideas of Epicurus, who was a follower of Democritus: Look closely, whenever rays are let in and pour the sun's light through the dark places in houses you will see many particles there stirred by unseen blows change their course and turn back, driven backwards on their path, now this way, now that, in every direction everywhere. You may know that this shifting movement comes to them all from the atoms.

For first the atoms of things move of themselves; then those bodies which are formed of a tiny union, and are, as it were, nearest to the powers of the atoms, are smitten and stirred by their unseen blows, and they, in their turn, rouse up bodies a little larger. And so the movement passes upwards from the atoms, and little by little comes forth to our senses, so that those bodies move too, which we can descry in the sun's light; yet it is not clearly seen by what blows they do it.

Is it possible some young Greeks had acute enough eyesight to see Brownian motion? These Greek philosophers believed that atoms were in constant motion, and always had been, at least in gases and liquids. Sometimes, however, as a result of their close-locking shapes, they joined in close-packed unions, forming materials such as rock or iron. Basically, Democritus and his followers had a very mechanical picture of the universe. They thought all natural phenomena could in principle be understood in terms of interacting, usually moving, atoms.

This left no room for gods to intervene. Their atomic picture included the mind and even the soul, which therefore did not survive death. This was in fact a cheerful alternative to the popular religions of the day, in which the gods constantly intervened, often in unpleasant ways, and death was to be dreaded because punishments would surely follow.

Little conceptual progress in atomic theory was made over the next two thousand years, in large part because Aristotle discredited it, and his views held sway through the Middle Ages.

GALILEO

Things began to look up with the Renaissance. Galileo believed in atoms, although, like the early Greeks, he seemed to confuse the idea of physical indivisibility with that of having zero spatial extent, i.e. being a mathematical point. Nevertheless, his ideas in this area apparently got him into theological hot water. The Church felt that the doctrine of transubstantiation - the belief that the bread and wine literally became the body and blood of Christ - was difficult to believe if everything was made up of atoms. This was an echo of the tension between atoms and religion two thousand years earlier.

Galileo's theory of atoms was not very well developed. He gives the impression in some places they were infinitely small, and in view of his excellent grasp of dimensional scaling arguments, he may have thought that vacuum suction between infinitesimally small surfaces would suffice to hold solids together, since smaller objects have proportionately more surface. Of course, this was on the wrong track. (Ironically, shortly after Galileo's death, his pupil Torricelli was the first to realise that suction forces were really a result of air pressure from the weight of the atmosphere.)

NEWTON

A much more modern perspective on atoms and interatomic forces was set out later in the seventeenth century by Isaac Newton, who wrote: Have not the small Particles of Bodies certain Powers, Virtues, or Forces, by which they act at a distance, not only upon the Rays of Light for reflecting, refracting and inflecting them, but also upon one another for producing a great Part of the Phenomena of Nature?

For it's well known, that Bodies act upon one another by the Attractions of Gravity, Magnetism, and Electricity; and these Instances show the Tenor and Course of Nature, and make it not improbable that there be more attractive Powers than these.... For we must learn from the Phenomena of Nature what Bodies attract one another, and what are the Laws and Properties of the Attraction, before we enquire the Cause by which the attraction is perform'd.

The Attractions of Gravity, Magnetism and Electricity, reach to very sensible distances, and so have been observed by vulgar eyes, and there may be others which reach to so small distances as hitherto escape Observation, and perhaps electrical Attraction may reach to such small distances, even without being excited by friction.

In fact, although the forces binding atoms together in molecules cannot be properly understood without quantum mechanics, many of these forces *are* "short range" electrical forces - forces between bodies having overall electrical neutrality, but distorted charge distributions. These forces could definitely be categorized as "electrical Attraction reaching to small distances". Notice that Newton also leaves the door open for *other* short range forces, which were finally discovered in the 1930's!

Newton goes on to argue that assuming the existence of forces of attraction between particles suggests very natural explanations for various physical chemistry-type phenomena, such as deliquescence, ease of distillation and heat of mixing:

For when Salt of Tartar runs per Deliquium, is not this done by an Attraction between the Particles of the Salt of Tartar, and the Particles of the Water which float in the Air in the form of Vapors? ... And whence is it but from this attractive Power that Water which alone distills with a gentle lukewarm heat, will not distill from Salt of Tartar without a great heat? ... And when Water and Oil of Vitriol poured successively into the same Vessel grow very hot in the mixing, does not this heat argue a great Motion in the Parts of the Liquors? And does not this Motion argue, that the Parts of the two Liquors in mixing coalesce with Violence, and by consequence rush towards one another with an accelerated Motion?

Evidently, Newton had already realized that heat is molecular motion, and how such heat is generated when dissimilar molecules that attract each other are mixed, so their potential energy translated into kinetic energy as they move towards each other.

PRE-CHEMISTRY: INCLUDING NEWTON THE ALCHEMIST

The impressive quotes from Newton above, which sound like he's on the right track, *do not tell the whole story*. Newton thought that part of chemistry (especially the physical part) *could* be explained in terms of the mechanics of corpuscles, but that there was something more important - a harder-to-pin-down vital spirit, which was the basis of life (and also somehow connected with mercury and other elements).

He also felt this was the key to the way God ran the universe the merely mechanical interaction of corpuscles could not, in his opinion, generate the rich variety of life. And Newton wanted to understand just how God *did* run the universe. Newton probably spent more time studying alchemy than he did working on his laws, gravitation and calculus combined!

In fact, Newton probed "the whole vast literature of the older alchemy as it has never been probed before or since" according to a recent historical study. He also used quite precise quantitative measures in many of his investigations. This did not provide the insight into mass conservation that Lavoisier's work did a century later, probably because Newton didn't count the various gases absorbed or emitted, these were still considered incidental and not really important to the reactions. Also, maybe they didn't smell too greata recipe for preparing phosphorus Newton copied from Boyle begins "Take of Urin one Barrel". Enough already.

Not that this matters too much as far as developing the atomic concept is concerned. On the positive side, the alchemists, in their fruitless quest to turn lead into gold (and find the elixir of life, etc.) did get very skillful at managing a great variety of chemical reactions, and so learned the properties of many substances.

The alchemists' point of view was based on Aristotle's four elements, earth, air, fire and water, but they added what they called principles. For example, there was an active principle in air important in respiration and combustion. There was an acidic principle, and others. And then there was phlogiston. Looking at something in flames, it seems pretty clear that something is escaping the material. That they called phlogiston. After Boyle discovered that metals become *heavier* on combustion, it was decided that phlogiston had negative weight.

LAVOISIER

The first major step towards modern quantitative chemistry was taken by Lavoisier towards the end of the eighteenth century. He realized that combustion was a chemical reaction between the material being burned and a component of the air. He carried out reactions in closed vessels so that he could keep track of the amounts of the various reagents involved.

One of his great discoveries was that in reactions, the total final weight of all the materials involved is exactly equal to the total initial weight. This was the first step on the road to thinking about chemistry in terms of atoms. He also established that pure water was not transmuted to earth by heating, as had long been believed - the residue left on boiling dry came from the container if the water itself was pure.

Lavoisier discovered oxygen. He was the first to realise that air has two (major) components, only one of which supports respiration, meaning life, and combustion. In 1783, working with the mathematician Laplace, and a guinea pig in a mask, he checked out quantitatively that the animal used breathed-in oxygen to form what we now term carbon dioxide (this is the origin of the "guinea pig" as experimental subject).

WHAT IS AN ELEMENT?

Lavoisier tightened up the very loose terminology in use at that time: there were no generally agreed on definitions of elements, principles or atoms, although a century earlier Boyle had suggested that element be reserved for substances that could not be further separated chemically.

In his Elements of Chemistry Lavoisier writes: If, by the term elements we mean to express those simple and indivisible atoms of which matter is composed, it is extremely probable that we know nothing about them; but if we apply the term elements, or principles of bodies, to express our idea of the last point which analysis is capable of reaching, we must admit as elements all the substances into which we are capable, by any means, to reduce bodies by decomposition.

Not that we are entitled to affirm that these substances we consider as simple may not be compounded of two, or even of a greater number of principles; but since these principles cannot be separated, or rather since we have not hitherto discovered the means of separating them, they act with regard to us as simple substances, and we ought never to suppose them compounded until experiment and observation have proved them to be so.

In sum, Lavoisier began the modern study of chemistry: he insisted on precise terminology and on precise measurement, and suggested as part of the agenda the classification of substances into *elements* and compounds. Once this programme was truly underway, the atomic interpretation soon appeared.

Unfortunately for chemistry, five years after this book appeared Lavoisier went to the guillotine. In pre-revolutionary France, government tax collection was privatized, and Lavoisier was one of the very unpopular "tax-farmers". Few of them survived the revolution. Lavoisier was also accused of anti-French activities, in that he corresponded with foreigners.

The fact that all the correspondence was exchange of

scientific papers did not impress the revolutionaries, who remarked that "the Republic has no need of savants" as they sent him to the guillotine.

DALTON

John Dalton (1766-1844) was born into a poor family near Manchester, England. He supported himself to some extent by teaching from the age of twelve, when he started his own small Quaker school. Dalton wrote A New System of Chemical Philosophy, from which the following quotes are taken:

Matter, though divisible in an extreme degree, is nevertheless not infinitely divisible. That is, there must be some point beyond which we cannot go in the division of matter. The existence of these ultimate particles of matter can scarcely be doubted, though they are probably much too small ever to be exhibited by microscopic improvements.

He assumed that all atoms of an element were identical, and atoms of one element could not be changed into atoms of another element "by any power we can control". He assumed further that compounds of elements had compound atoms:

Now, though this atom may be divided, yet it ceases to be carbonic acid, being resolved by such division into charcoal and oxygen. He also asserted that all compound atoms (molecules, as we would say) for a particular compound were identical, and, furthermore: "Chemical analysis and synthesis go no farther than to the separation of particles one from another, and to their reunion. No creation or destruction of matter is within reach of chemical agency".

By Dalton's time it had become clear that when elements combine to form a particular compound, they always do in precisely the same ratio by weight. For example, when hydrogen burns in oxygen to form water, one gram of hydrogen combines with eight grams of oxygen. This constancy is to be expected in Dalton's theory, presumably the compound atom, or molecule, of water has a fixed number of hydrogen atoms and a fixed number of oxygen atoms.

Of course, the weight ratio doesn't tell us the numbers, since we don't know the relative weights of the hydrogen atom and the oxygen atom. To make any progress, some assumptions are necessary. Dalton suggested a rule of greatest simplicity: if two elements form only one compound, assume the compound atom has only one atom of each element.

Since H_2O_2 had not been discovered, he assumed water was HO. (He actually used symbols to represent the elements, H was a circle with a dot in the centre. However, just as we do, he used strings of such symbols to represent an actual molecule, not a macroscopic mixture.) On putting together data on many different reactions, it became apparent to Dalton that the rule of greatest simplicity wasn't necessarily correct, by 1810 he was suggesting that the water molecule perhaps contained three atoms.

DALTON'S MULTIPLE PROPORTIONS

One of the strongest arguments for Dalton's atomic theory of chemistry was the Law of Multiple Proportions. For example, he found that when carbon combined with oxygen to form a gas, there were two possible outcomes, depending on the conditions - and in one outcome each gram of carbon combined with precisely twice as much oxygen as in the other. He correctly interpreted this as the formation of CO_2 and CO respectively.

GAY-LUSSAC'S SIMPLE RATIOS: AND A BALLOON RIDE

Meanwhile in Paris, Joseph Louis Gay-Lussac investigated carefully the ratio of the volume of hydrogen gas that combined with a given volume of oxygen gas to form water. He found the oxygen could combine with exactly twice its own volume of hydrogen.

There were similar simple volumetric ratios for other reactions between gases, and furthermore, if the product of the reaction was also a gas, it filled a volume simply related to those of the combining gases - so two volumes of hydrogen combined with one volume of oxygen to give two volumes of steam (assuming of course the temperature is not allowed to cool below the boiling point of water).

Unfortunately, Dalton didn't believe Gay-Lussac's results. Dalton was convinced that Newton had proved the atoms of a gas were large, elastic objects, essentially filling space, and of different sizes for different atoms. This was hard to reconcile with the simple volume ratios.

Gay-Lussac had guts. Since it was now clear that nitrogen was a little lighter than oxygen, he thought there might be proportionately less oxygen in the air at higher elevations. To find out, in 1802 he went up in a balloon to 23,000 feet! He found the mix to be pretty much the same.

AVOGADRO'S HYPOTHESIS

In 1811, the Italian physicist Amedeo Avogadro suggested that Dalton's picture of atoms and molecules could be reconciled with Gay-Lussac's results on volumes if one assumed that equal volumes of all gases, elements or compounds, contain equal numbers of molecules. Of course, he had no idea what the number might be, but the hypothesis made many predictions without knowing the number.

Dalton didn't buy this one either. For one thing, if one volume of oxygen combined with hydrogen to form two volumes of water, it looked to Dalton as if the water molecules each had half an oxygen atom, if we believe Avogadro's hypothesis. Dalton did not believe that the original oxygen gas could consist of diatomic molecules, because in his picture the large oxygen atoms repelled each other, that's why a gas resisted compression. So how could they attract each other to form molecules?

RETURN OF THE KINETIC THEORY

Although Dalton's work had set the agenda for chemistry and led to many fruitful investigations, his picture of a gas was a roadblock to a real understanding of gas reactions. The large molecules he envisioned had small centers surrounded by an atmosphere of caloric, which was a fluid of heat, the same stuff you feel seeping into your fingers when you touch something hot. The reason a gas expanded on heating, in this theory, was that the caloric moved in and attached itself to the atmospheres around the molecules (Lavoisier also believed in caloric).

However, the work of Rumford and later physicists established that heat was best understood as molecular motion, and there was no fluid - the caloric was just an illusion. This made the whole Newton/Dalton picture difficult to believe. In spite of this, Bernoulli's 1738 kinetic model, in which tiny gas molecules were shooting around in otherwise empty space, was not widely debated until *half a century* after Avogadro's hypothesis. At that point, our modern picture of gases began to emerge - the results of Dalton, Gay-Lussac and Avogadro could be put together in a simple, consistent way.

ELECTRICALLY PRODUCED ELEMENTS

Volta invented the electric battery in 1800, and it was used within weeks to electrolyze water into its constituent elements, hydrogen and oxygen. More surprisingly, when an electric current was passed through soda or potash - two substances previously thought to be elements - metallic substances appeared at the cathodes. New elements-sodium and potassium - had been discovered. Evidently, electrolysis could break up compounds that were impervious to chemical attack.

Chlorine gas was discovered by Sir Humphry Davy, on electrolyzing muriatic acid. His assistant Michael Faraday went on to analyse electrolysis quantitatively. He found that when electrolysis liberated elements at an electrode, it took always the same total amount of electric current (or some small integer multiple) to liberate one mole of the element (that is, Avogadro's number of atoms).

The real significance of this result was not fully realized until 1881, when Helmholtz, in a Faraday Memorial Lecture, said: "If we accept the hypothesis that elementary substances are composed of atoms, we cannot well avoid concluding that electricity also is divided into elementary portions which behave like atoms of electricity".

PATTERNS OF ELEMENTS

In the year 1800, 31 elements were known. By 1860, that number had almost doubled, to 60, and the relative atomic weights, as well as many of the chemical properties, were known. In particular, in analyzing molecule formation, a valuable emerging concept was that of *valence* - the idea that each atom had a particular number of little hooks on it to attach itself to similar hooks on other atoms. Some atoms exhibited different valencies in different compounds, but many didn't, so it was a useful guide. In 1865, the English chemist J. A. R. Newlands looked for correlations between chemical properties, including valence, and atomic weight The lightest elements known at the time, in order of increasing atomic weight, were:

H, Li, Be, B, C, N, O, F, Na, Mg, Al, Si, P, S, Cl, K, Ca, Ti, ...

Newlands suggested there was a *law of octaves* - elements 1, 8, 15 were a lot alike, as were 2, 9, 16, and so on. (Hydrogen is sufficiently anomalous that it does not fit in the pattern as convincingly as the other light elements.) Newlands' colleagues were unconvinced. For one thing, the pattern did not seem to extend much further up the list. One colleague suggested to Newland that he search for patterns by putting the list in alphabetical order, rather than by atomic weight.

A few years later, in 1872, the Russian physicist Mendeleev drew up his periodic Table of the Elements, with eight columns reflecting Newland's octaves. Mendeleev had the courage to leave gaps where he believed new elements would be discovered - so that the periodic patterns of chemical behaviour recurred throughout the table. For a specific example, he predicted an element of atomic weight about 72, to follow silicon in his table.

He gave a list of expected properties of its compounds, such as a tetrachloride of density 1.9 and boiling point about 90 degrees, etc. This prediction was made in 1871, and in 1887 germanium was discovered, atomic weight 72.5, the tetrachloride had density 1.9 and boiled at 83 degrees. His predictions for other compounds of germanium were similarly accurate.

These discoveries led to acceptance of the validity of Mendeleev's work towards the end of the century. The one big surprise was the inert gases He, Ne, Ar, Kr, Xe, Rn - a whole new column was needed! Mendeleev was pretty upset by this development, but accepted it in the end.

Everybody Didn't Buy it

It is worth bearing in mind that even as late as the 1890's

some of the most eminent German chemists did not accept the atomic idea. Here is an 1895 quote from Ostwald: "The proposition that all natural phenomena can ultimately be reduced to mechanical ones cannot even be taken as a useful working hypothesis: it is simply a mistake. This mistake is most clearly revealed by the following fact.

All the equations of mechanics have the property that they admit of sign inversion in the temporal quantities. That is to say, mechanical processes can develop equally well forwards or backwards in time. Thus, in a purely mechanical world there could not be a before and an after as we have in our world: the tree could become a shoot and a seed again .... The actual irreversibility of natural phenomena thus proves the existence of processes that cannot be described by mechanical equations.

Of course, this argument harks back to the kinetic theory, entropy and Boltzmann, Ostwald was one of the reasons Boltzmann committed suicide. Nevertheless, within few years, Boltzmann's ideas were widely accepted. The atom had finally come to stay.

ATOMIC SPECTRA

The first person to realise that white light was made up of the colors of the rainbow was Isaac Newton, who in 1666 passed sunlight through a narrow slit, then a prism, to project the colored spectrum on to a wall. This effect had been noticed previously, of course, not least in the sky, but previous attempts to explain it, by Descartes and others, had suggested that the white light became colored when it was refracted, the colour depending on the angle of refraction.

Newton clarified the situation by using a second prism to reconstitute the white light, making much more plausible the idea that the white light was composed of the separate colors. He then took a monochromatic component from the spectrum generated by one prism and passed it through a second prism, establishing that no further colors were generated. That is, light of a single colour did not change colour on refraction.

He concluded that white light was made up of all the colors of the rainbow, and that on passing through a prism, these different colors were refracted through slightly different angles, thus separating them into the observed spectrum.

In 1752, the Scottish physicist Thomas Melvill discovered that putting different substances in flames, and passing the light through a prism, gave differently patterned spectra. Ordinary table salt, for example, generated a "bright yellow". Furthermore, not all the colors of the rainbow appeared - there were dark gaps in the spectrum, in fact for some materials there were just a few patches of light. By the 1820's, Herschel had recognized that spectra provided an excellent way to detect and identify small quantities of an element in a powder put into a flame.

Meanwhile, the white light of the sun was coming in for more detailed scrutiny. In 1802, William Wollaston in England had discovered (perhaps by using a thinner slit or a better prism) that in fact the solar spectrum itself had tiny gaps - there were many thin dark lines in the rainbow of colors. These were investigated much more systematically by Joseph von Fraunhofer, beginning in 1814. He increased the dispersion by using more than one prism. He found an "almost countless number" of lines. He labeled the strongest dark lines A, B, C, D, etc.

FOUCAULT CONNECTS MELVILL'S BRIGHT LINES AND FRAUNHOFER'S DARK LINES

In 1849, Foucault (of speed of light and pendulum fame) examined the spectrum of light from a voltaic arc between carbon poles. He saw a bright double yellow line at exactly the same wavelength as Fraunhofer's dark D line in the solar spectrum. Investigating further, Foucault passed the sun's light through the arc, then through a prism. He observed that the D lines in the spectrum were even darker than usual. After testing with other sources, he concluded that the arc, which emitted light at the D line frequency, would also absorb light from another source at that frequency.

This discovery did not surprise Sir George Stokes in Cambridge. He pointed out that any mechanical system with a natural frequency of oscillation will emit at that frequency if disturbed, but will also absorb most readily at that frequency from incoming disturbances, the phenomenon of resonance.

Question: In a total eclipse of the sun, the only sunlight reaching earth comes from the hot gases of the sun's atmosphere, light from the sun's main disc being blocked by our moon. The light from these hot gases was analyzed during an eclipse in 1870. How do you think the spectrum observed related to that of full sunlight?

The spectrum of *hydrogen*, which turned out to be crucial in providing the first insight into atomic structure over half a century later, was first observed by Anders Angstrom in Uppsala, Sweden, in 1853. His communication was translated into English in 1855. Angstrom, the son of a country minister, was a reserved person, not interested in the social life that centered around the court. Consequently, it was many years before his achievements were recognized, at home or abroad (most of his results were published in Swedish).

Meanwhile, in Freeport, Pennsylvania, in 1855, David Alter described the spectrum of hydrogen and other gases. In the 1840's, Alter had started the first commercial production of bromine from brines. He also found a way to extract oil from coal, but that proved uneconomic after the discovery of oil in Pennsylvania. His work was not widely recognized, either.

BUNSEN AND KIRCHHOFF

The first really systematic investigation of spectra was that of Bunsen and Kirchhoff, in Heidelberg, between 1855 and 1863. They used several techniques. For one thing, they introduced various salts intowhat else?the flame of a Bunsen burner. This was a very effective way of viewing spectra, because the Bunsen burner flame itself gave out practically no light. They also used the cooler flame of alcohol burning mixed with water to generate a vapour to study absorption spectra.

Finally, they studied the spectra of electric arcs between electrodes of different materials. Using iron electrodes gave a spectrum that coincided with dark lines in the sun's spectrum. Copper electrodes did not. They concluded that the sun's atmosphere contained iron, but not much copper, and that, they said, seemed very plausible since there is so much iron in the earth, and in meteors.

(*Cautionary note to philosophers*: In 1835, the French philosopher Auguste Comte (the founder of positivism) wrote: "...Our knowledge concerning the gaseous envelopes [of stars] is necessarily limited to their existence, size and refractive power, we shall not at all be able to determine their chemical composition or even their density.

The collaboration of Kirchhoff and Bunsen was a major research effort, even by modern standards. They determined thousands of spectral lines, each to an accuracy of one part in ten thousand. They spectroscopically discovered new elements: rubidium and cesium.

Their method was used to find fifteen more new elements before the end of the century. In 1869, Joseph Lockyer studied the spectra of solar prominences (in eclipses). He found the spectra to be slightly Doppler shifted, so was able to deduce the speeds of the gases whirling around the sunspots. He also found a spectrum never seen before, and conjectured that it came from a new element he named Helium.

In fact, helium was later discovered on earth in 1895, by Ramsay. At that time, it had just become evident that there was an inert component, argon, in the earth's atmosphere. Earlier, an inert gas had been observed to emanate from uranium salts when they were heated. Ramsay assumed this would be the same gas, but decided to check. On heating uranium salts and performing a spectral analysis of the emitted gas, much to his surprise he found it to be helium.

THE BALMER SERIES

It is clear from the above that a tremendous amount of scientific progress was made using spectral lines, yet no-one had the slightest idea why atoms emitted at the frequencies they did. It was appreciated that spectra implied that atoms had structure.

In 1852, Stokes had stated that probably the vibrations that produced light were vibrations among the constituent parts of molecules (a term which also included atoms at that time) and in 1875 Maxwell, in enumerating properties atoms must have, included the capability of internal motion or vibration. This worried Maxwell, though. As he said, the spectroscopic evidence forces the conclusion that the atom is quite complex, with many internal degrees of freedom. Yet apparently all these modes of vibration, or almost all of them, are not excited by heat, since if they were this extra capacity of the atom to absorb energy would be reflected in its specific heat.

Obviously, if any pattern could be discerned in the spectral lines for an atom, that might be a clue as to the internal structure of the atom. One might be able to build a model. A great deal of effort went into analyzing the spectral data from the 1860's on. The big breakthrough was made by Johann Balmer, a math and Latin teacher at a girls' school in Basel, Switzerland. Balmer had done no physics before, and made his great discovery when he was almost sixty.

He decided that the most likely atom to show simple spectral patterns was the lightest atom, hydrogen. Angstrom had measured the four visible spectral lines to have wavelengths 6562.10, 4860.74, 4340.1 and 4101.2 in Angstrom units (10^{-10} meters). Balmer concentrated on just these four numbers, and found they were given by the formula:

$$\lambda = b\left(\frac{n^2}{n^2 - 4}\right)$$

where $b = 3645.6$ Angstroms, and $n = 3, 4, 5, 6$. Balmer suggested that there would be other lines - in the infrared - corresponding to $n = 7, 8$, etc., and in fact some of them had already been observed, unbeknownst to Balmer. He further conjectured that the 4 could be replaced by 9, 16, 25, ... and this also turned out to be true - but these lines, further into the infrared, were not detected until the early twentieth century, along with the ultraviolet lines generated by replacing the 4 by 1.

It is instructive to write Balmer's general formula in terms of the inverse wavelength. This is called the wave number - the number of waves that fit in one unit of length.

$$\frac{1}{\lambda} = R_H = \left(\frac{1}{n^2} - \frac{1}{m^2}\right)$$

where n, m are integers, and R_H is the Rydberg constant, 109,737 cm^{-1}.

This constant is named after the Swedish physicist Rydberg who (in 1888) presented a generalization of Balmer's formula, in which the integer n was replaced by n + constant, the constant being less than unity. Rydberg suggested that all atomic spectra formed families with this pattern. (He also said he was unaware of Balmer's work.) It turns out that there *are* families of spectra following Rydberg's pattern, notably in the alkali metals, sodium, potassium, etc., but not with the precision the hydrogen atom lines fit the Balmer formula, and low values of n give lines that deviate considerably.

(Modern footnote: atoms having spectral lines following Rydberg's formula are called Rydberg atoms. These Rydberg atoms have one electron orbiting at a much greater distance from the nucleus than the others. Consequently, Rydberg atoms can only survive in a gas at very low pressure, otherwise that outermost electron gets knocked off. Prof. Tom Gallagher in our Department is a world expert on these atoms, which have proven a rich source of information on the quantum mechanics of atomic structure.)

One pattern that was noticed in the spectra of many atoms is Ritz' Combination Principle: if for a given atom there are spectral lines at two wave numbers, there is sometimes another spectral line at the precise sum of those two wave numbers.It is easy to see from Balmer's formula that this is true for some pairs of lines in the hydrogen spectrum. It also turns out to be true for atoms where the spectral lines have no other discernible pattern.

MODELS OF THE ATOM

The first attempt to construct a physical model of an atom was made by William Thomson (later elevated to Lord Kelvin) in 1867. The most striking property of the atom was its *permanence*. It was difficult to imagine any small solid entity that could not be broken, given the right force, temperature or chemical reaction. In contemplating what kinds of physical systems exhibited permanence, Thomson was inspired by a paper Helmholtz had written in 1858 on *vortices*.

This work had been translated into English by a Scotsman, Peter Tait, who showed Thomson some ingenious experiments with smoke rings to illustrate Helmholtz' ideas. The main point was that in an *ideal* fluid, a vortex line is always composed of the same particles, it remains *unbroken*, so it is ring-like. Vortices can also form interesting combinationsA good demonstration is provided by creating two vortex rings one right after the other going in the same direction.

They can trap each other, each going through the other in succession. This is probably what Tait showed Thomson, and it gave Thomson the idea that atoms might somehow be vortices in the ether.

Of course, in a non ideal fluid like air, the vortices dissipate after a while, so Helholtz' mathematical theorem about their permanence is only approximate. But Thomson was excited because the ether *was* thought an ideal fluid, so vortices in the ether might last forever!

This was very aesthetically appealing to everybody - "Kirchhoff, a man of cold temperament, can be roused to enthusiasm when speaking of it." In fact, the investigations of vortices, trying to match their properties with those of atoms, led to a much better understanding of the hydrodynamics of vortices - the constancy of the circulation around a vortex, for example, is known as Kelvin's law.

In 1882 another Thomson, J. J., won a prize for an essay on vortex atoms, and how they might interact chemically. After that, though, interest began to wane - Kelvin himself began to doubt that his model really had much to do with atoms, and when the electron was discovered by J. J. in 1897, and was clearly a component of all atoms, different kinds of non-vortex atomic models evolved.

It is fascinating to note that the most exciting theory of fundamental particles at the present time, *string* theory, has a definite resemblance to Thomson's vortex atoms. One of the basic entities is the closed string, a little loop, which has fields flowing around it reminiscent of the swirl of ethereal fluid in Thomson's atom. And it's a very beautiful theory - Kirchhoff would have been enthusiastic!

FLOATING MAGNETS

In 1878, Alfred Mayer, at the University of Maryland,

dreamed up a neat demonstration of how he imagined atoms might be arranged in molecules. He took a few equally magnetized needles and stuck them through corks so that they would float with their north poles all at the same height above the water, all repelling each other equally.

He then held the south pole of a more powerful magnet some distance above the water, to attract the needles towards this central point. The idea was to see what equilibrium patterns the needles would form for different numbers of needles. He found something remarkable - the needles liked to arrange themselves in shells. Three to five magnets just formed a triangle, square and pentagon in succession. but for six magnets, one went to the centre and the others formed a pentagon. For more magnets, an outer shell began to form.

Kelvin's immediate response to Mayer's publication was that this should give some clues about the vortex atom. Apparently it didn't, but twenty-five years later it guided his thinking on a new model.

PLUM PUDDING

Kelvin, in 1903, proposed that the atom have the newly discovered electrons embedded somehow in a sphere of uniform positive charge, this sphere being the full size of the atom. (Of course, the sphere itself must be held together by unknown non-electrical forces - which is still true of the positive charge in our modern model of the atom.)

This picture was taken up by J. J. Thomson too, and was dubbed the plum pudding model, after traditional English Christmas fare, a large round pudding (rich with suet) with raisins embedded in it. In 1906, J. J. concluded from an analysis of the scattering of X-rays by gases and of absorption of beta-rays by solids, both of which he assumed were effected by electrons, that the number of electrons in an atom was approximately equal to the atomic number.

This led to a picture of electron arrangements in an atom reminiscent of Mayer's magnets. Perhaps by analyzing possible modes of vibration of electrons in these configurations, the spectra could be calculated.

The simplest case to consider was clearly hydrogen, now assumed (correctly) to contain just one electron.

How Does an Atom's Colour Depend on its Size?

By "colour" we mean here the spectral colors emitted when the atom is excited. In Thomson's plum pudding model, there is a clear relationship between the *size* of the pudding and the *frequency* at which the electron will oscillate, and hence presumably radiate, when excited.

The two are related because the assumption is that the total positive charge - which is uniformly spread throughout the sphere is just equal to the electron's negative charge. At rest in its lowest state, the electron just sits in the middle of this sphere of charge. When bumped somehow, it will oscillate about that point. If the electron is at distance x from the centre, it will feel a restoring force towards the centre equal to the attraction from that part of the positive charge it is "outside" of - that is, the charge within a sphere of radius x about the centre.

Therefore, the *larger* the whole atomthe pudding - the more thinly spread the positive charge is, and the *smaller* the amount of charge within the small sphere of radius x that is attracting the electron back towards the centre. So, the bigger the atom is, the slower the electron's oscillation is, and the lower frequency the radiation emitted.

It is straightforward to give a quantitative estimate of the size of the atom based on the observation that when excited it emits radiation in the visible range. Let us assume that the positively charged sphere has radius r_0 (this is then the size of the atom, which we know is about 10^{-10} meters).

If the electron is displaced from the centre of the atom in the x-direction an amount x, it is attracted back by all the charge that is now closer to the centre than itself, that is, an amount of charge equal to ex^3/r_0^3. (Recall e is the total amount of charge on the sphere, and x^3/r_0^3 is the fraction of the sphere closer to the centre than x.)

This charge acts as if it were a point charge at the origin, so the inverse-square law gives a $1/x^2$ factor, and the equation of motion for the electron is therefore:

$$m\frac{d^2x}{dt^2} = -\frac{1}{4\pi\varepsilon_0}\cdot\frac{e^2x}{r_0^3}$$

Provided it stays within the sphere, the electron will execute simple harmonic motion with a frequency

$$\omega^2 = \frac{1}{4\pi\varepsilon_0}\cdot\frac{e^2}{mr_0^3}.$$

Notice that, as we discussed above, as the size of the atom increases the frequency goes down. And we know the frequency, at least approximately it corresponds to visible light. Therefore, this model will predict a size of the atom, which we can compare with the size from other predictions, such as Brownian motion (plus the assumption that in a liquid, the atoms are fairly close packed - they take up most of the room available).

If we take visible light, say with a frequency 4.10^{15} radians per second, we find r_0 must be about 2.10^{-10} meters, a little on the large side, but encouragingly close to the right answer for a first attempt.

Sad to report, though, no real progress was made beyond this in predicting spectra using Thomson's pudding. Many attempts were made to find stable arrangements of electrons in atoms, not just hydrogen, using models like Mayer's magnets, and also having the electrons going around in circles. It was hoped that if certain numbers of magnets formed a very stable arrangement, that might model a chemically nonreactive atom, etc. - but nobody succeeded in making any real predictions along these lines, the models could not be connected with the properties of real atoms.

Evidently, then, the theorists were stuck - and the experimental challenge was to find some way to look *inside* an atom, and see how the electrons were arranged.

THE BOHR ATOM

In 1911, the 26-year-old Niels Bohr earned a Ph. D. at the University of Copenhagen; his dissertation was titled "Studies on the Electron Theory of Metals". He was awarded a postdoctoral fellowship funded by the Carlsberg Brewery Foundation, which enabled him to go to Cambridge in September to study with J. J.

Thomson. Bohr was a great admirer of Thomson's many achievements, both experimental and theoretical. In his thesis work, he had closely studied some of the problems covered in Thomson's book *Conduction of Electricity through Gases.*

He had uncovered some apparent errors in Thomson's work, and looked forward to discussing these points with the great man. Unfortunately, by the time Bohr arrived, the Cavendish Laboratory had grown to the point where Thomson as director had more than he could manage.

He had no spare time to think about electrons, and was not happy to hear from Bohr that some of his earlier work might be incorrect. In fact, Thomson went out of his way to avoid theoretical discussions with Bohr. He did assign Bohr an experiment on positive rays, but Bohr was not enthusiastic. Bohr kept himself busy writing a paper on electrons in metals, reading Dickens to improve his English, and playing soccer.

In December, Rutherford came down from Manchester for the annual Cavendish dinner. Bohr later said that he was deeply impressed by Rutherford's charm, his force of personality, and his patience to listen to every young man who might have an ideacertainly a refreshing change after J. J.! A little later, Bohr met with Rutherford again when he visited one of his father's friends in Manchester, someone who also knew Rutherford.

Although Rutherford was usually skeptical of theorists, he liked Bohr. For one thing, Rutherford was a soccer fan, and Bohr's brother Harald (only nineteen months younger than Bohr) was famoushe had played in the silver medal winning Danish soccer team at the 1908 Olympics in London. After talking it over with Harald, who visited Cambridge in January, Bohr moved to Manchester in March, and took a six-week lab course, given by Geiger, Marsden and others. Really, though, his interests were theoretical, and he talked a lot with Charles Galton Darwin"grandson of the real Darwin", as Bohr put it in a letter to Harald.

Darwin had just completed a theoretical analysis of the loss of energy of an a -particle going through matterthat is, an a that doesn't get close enough to a nucleus to be scattered. Such a 's gradually lose energy by churning through the electrons, and the rate of loss depends on how many electrons they encounter. In particular, Bohr concluded, after reviewing and improving on Darwin's work, it seemed clear that the hydrogen atom almost certainly had a *single* electron outside the nucleus.

WHAT DETERMINES ATOMIC SIZE?

A big problem with the nuclear hydrogen atom was: what determined its size? Classical mechanics gives a simple dynamical equation for circular orbits:

$$\frac{mv^2}{r} = \frac{1}{4\pi\varepsilon_0}\cdot\frac{e^2}{r^2}.$$

Now this equation is satisfied by *any* circular orbit centered at the nucleus, however large or small. (Note, by the way, that multiplying both sides by $r/2$ gives that the magnitude of the kinetic energy in the circular orbit is just half the magnitude of the negative potential energy. We need this below.) There is no hint here that the atom in its "natural" ground state should have any particular radius. But it does! This means we're missing *something*. But what?

Bohr (and others) thought that Planck's constant must somehow play a role in determining the size of the orbit. After all, it *did* play a role in restricting allowed orbital changes in the oscillators in black body radiationand these oscillators, although not very clearly understood, were of the same general size as atoms.

So evidently the standard picture of how an oscillating charge radiated couldn't be right at the atomic level. Bohr concluded that in an atom in its natural rest state, the electron must be in a special orbit, he called it a "stationary state" to which the usual rules of electromagnetic radiation didn't apply. In this orbit, which determined the size of the atom, the electron, mysteriously, didn't radiate.

Just how to bring Planck's constant into a discussion of the hydrogen atom was not so clear, though. For the black body oscillators, it related the frequency f of the oscillator with the allowed energy change E by $E = hf$. The obvious parallel approach for the hydrogen atom was to identify the frequency f with the circular frequency of the electron in its orbit. However, in contrast to the simple harmonic oscillator this hydrogen atom frequency *varied* with the size of the orbit. Still, it was the only frequency around, and, dimensionally, multiplying it by h gave an energy. What energy could that be identified with? Again, the choice was limitedthe electron had a kinetic energy E, the potential energy was $-2E$ and the total energy $-E$. If a hydrogen nucleus captured a passing electron into its ground state, and emitted one quantum of electromagnetic radiation, that quantum would have energy E, the same as the electron kinetic energy in the natural stationary state (called the *ground state*).

Bohr suggested in a note to Rutherford in the summer of 1912 that requiring this energy be some constant (assumed to be of order of magnitude one) multiplied by hf would fix the size of the atom. Actually his argument was a bit more complicated, he considered the several electron atom, and took the electrons to form rings. However, the basic point is the samea condition like this constrains the atomic size, it would be fixed uniquely if we knew the constant. If we assume the constant is 1, for example, we have

$$\frac{1}{2}mv^2 = \mathrm{hv}/2\pi\mathrm{r}$$

Putting this together with the dynamic equation above determines the atomic radius r. It is easy to check that it predicts a radius of 4p e $_0h^2$/p $^2me^2$, which is just four times the "right answer" defined as the Bohr radius. The correct Bohr radius comes out if we choose the constant to be one-half, $E = \frac{1}{2}hf$, which Bohr used later. Hence the approximate size of the atom follows from *dimensional* arguments alone once one assumes that Planck's constant plays a role! Of course, the nucleus is irrelevant in determining the atomic sizeit just provides a fixed centre of electrostatic attraction. The relevant electronic parameters are the mass m and the strength of attraction e^2/4p e $_0$. Together with h, these parameters determine a length.

It should be mentioned that this assumption explained more than the size of the hydrogen atom. It was believed at the time that in the higher atoms, the electrons formed rings, thought to lie one outside the other, and various stability arguments indicated that there couldn't be more than seven electrons in a ring. The length scale above, $4\pi\varepsilon_0 h^2/2\pi^2 me^2$, would *decrease* for larger atoms, with e^2 replaced by Ze^2 essentially, for nuclear charge Z. Thus as the number of rings increased, the size of the rings would decrease, explaining the observed approximate periodicity in atomic volume with atomic number.

Also, in 1911 Richard Whiddington in Cambridge had found that to cause a substance having atomic number A to emit characteristic x-rays by bombarding it with electrons, it was necessary to use electrons of speed approximately $A \times 10^6$ meters per second. Any substance on being bombarded with sufficiently fast electrons emits a continuum of x-ray frequencies up to a maximum frequency f given by hf = kinetic energy of electron, *plus* some sharply defined linesx-rays at a particular frequency, which does not change as the electron speed is further increased.

The frequency corresponding to these lines was found to increase with atomic number. Applying his length scale argument to the innermost ring of an atom, Bohr found that an electron in that ring would have a speed proportional to the nuclear charge, and hence, at least approximately, to the atomic number. Furthermore, the predicted speed in orbit was of the same order as that of Whiddington's electrons.

NICOLSON: A CLEVER IDEA ABOUT A WRONG MODEL

Meanwhile, in Cambridge one J. W. Nicolson was struggling to incorporate Planck's ideas in a model of the atom, in an attempt to understand some strange sets of spectral lines observed in nebulae and in the sun's corona. He conceived a rather exotic (and quite wrong!) model, in which a ring of electrons, like a necklace, orbited the nucleus. (Actually, many people, including Bohr himself, investigated models like this.

The reason was that the classical radiation from a ring of electrons is a lot less that that from a single orbiting electron, the fields tend to cancel each other.) Oscillations of electrons in this ring gave the spectra. Nicolson predicted the frequencies emitted by a straightforward classical analysis of these oscillation frequencies, in the spirit of earlier work on the plum pudding model.

He did bring in Planck's constant, though. He knew that

dimensionally it was a unit of angular momentum, and he suggested that the atom could only lose angular momentum in discrete amountspresumably constant multiples of h. Nicolson felt that, given the dimensionality of Planck's constant, quantization of angular momentum was more plausible than quantization of energy. Of course, for the simple harmonic oscillator they amounted to the same thing, but not for any other system.

BOHR RETURNS TO DENMARK

Bohr left Manchester in July 1912 and was married on the first of August. In the fall, he began work at the University of Copenhagen. At the same time, he began setting down on paper some of his Manchester ideas about atoms. He read Nicolson's work. As he wrote to Rutherford at the end of January 1913, he and Nicolson were really looking at different thingsNicolson was considering atoms in a very hot environment (like the sun's corona, or an electrical discharge tube) and the spectra gave information about how energy was emitted as the atom settled into its ground state.

Bohr himself was only interested in the state in which the system possessed the smallest amount of energy. He went on: " The question of calculation of the frequencies corresponding to the visible part of the spectrum". At that time, Bohr thought of spectra as pretty but peripheral, having as little to do with basic physics as the colors of a butterfly had to do with basic biology.

BOHR CHANGES HIS MIND ABOUT SPECTRA

In February 1913, Bohr was surprised to find out in a casual conversation with the spectroscopist H. R. Hansen that some patterns had been discerned in the apparent chaos of spectral lines. In particular, Hansen (a colleague and former classmate of Bohr) showed him Balmer's formula for hydrogen. They had very likely seen this in class together, but, given Bohr's opinion of the value of spectra, he probably hadn't paid much attention.

Balmer's formula is:

$$\frac{1}{\lambda} = R_H\left(\frac{1}{4} - \frac{1}{n^2}\right)$$

for the sequence of wavelengths of light emitted, with $n = 3, 4, 5, 6$ being in the visible, the lines used by Balmer in finding the formula. Hansen would doubtless have informed Bohr that the 1/4 could be replaced by $1/m^2$, with m another integer. The constant appearing on the right hand side is called the *Rydberg constant*, $R_H = 109{,}737\ \text{cm}^{-1}$. (This is the modern value-Balmer got it right to one part in 10,000, about the limit of spectral measurements at the time.)

Bohr said later: "As soon as I saw Balmer's formula, the whole thing was immediately clear to me." What he saw was that the set of allowed *frequencies* (proportional to inverse wavelengths) emitted by the hydrogen atom could all be expressed as *differences*.

This immediately suggested to him a generalization of his idea of a "stationary state" lowest energy level, in which the electron did not radiate. There must be a *whole sequence* of these stationary states, with radiation only taking place as the atom jumps from one to another of lower energy, emitting a single quantum of frequency f such that

$$hf = E_n - E_m,$$

the difference between the energies of the two states. Evidently, from the Balmer formula and its extension to general integers m, n, these allowed non-radiating orbits, the stationary states, could be labeled 1, 2, 3,..., n,... and had energies $-1, -1/4, -1/9, \ldots, -1/n^2, \ldots$ in units of hcR_H (using $\lambda f = c$ and the Balmer equation above). The energies are of course negative, because these are bound states, and we count energy zero from where the two particles are infinitely far apart.

Bohr was very familiar with the dynamics of simple circular orbits in an inverse square field. He knew that if the energy of the orbit was $-hcR_H/n^2$, that meant the kinetic energy of the electron, $\frac{1}{2}mv^2 = hcR_H/n^2$, and the potential energy would be.

$$-(1/4\text{pe}_0)e^2/r = -2hcR_H/n^2.$$

It immediately follows that the *radius* of the nth orbit is proportional to n^2, and the *speed* in that orbit is proportional to $1/n$. It then follows that the angular momentum of the nth orbit is just proportional to n.

Evidently, then the angular momentum in the nth orbit was *nKh*, where *h* is Planck's constant and *K* is some multiplying factor, the same for all the orbits, still to be determined. In fact, the value of *K* follows from the results above. R_H, *m*, *h*, and *c* are all known quantities (R_H being measured experimentally by observing the lines in the Balmer series) so the above formulas immediately give the electron's speed and distance from the nucleus in the nth orbit, and hence its angular momentum. Therefore, by putting in these experimentally determined quantities, we can find *K*.

BOHR FINDS THE RYDBERG CONSTANT WITHOUT DOING AN EXPERIMENT

The Balmer formula gave Bohr the essential clue that led to the realization that the angular momentum was quantized: $L = Kh$, $2Kh$, $3Kh$,... where *h* is Planck's constant, as usual, and *K* is some constant numerical factor, presumably of order 1.

Bohr gave a very clever argument to find K without doing any experiment. First, think about how the size of K affects the *physical* properties of the hydrogen atom. How would the atom be different for $K = 10$ compared with $K = 1$? For $K = 1$, the allowed orbits would be those having angular momentum *h*, 2*h*, 3*h*, 4*h*,.... For $K = 10$, the only allowed orbits would be those having angular momentum 10*h*, 20*h*,.... Evidently, for $K = 10$ there will be a lot fewer spectral lines, and the average *spacing* between them will be *greater* that for $K = 1$.

Next, Bohr imagined a really immense hydrogen atom, an electron going around a proton in a circle of one meter radius, say. This would have to be done in the depths of space, but really this is just a thought experiment in the spirit of Einstein. The point is that for this very large atom, the electron is moving rather slowly over a distance scale we are familiar with.

We know from many experiments that charges moving at these slow speeds over ordinary (human size) distances emit radiation according to Maxwell's equations. Or, more simply, if it's going round the circle at frequency *f* revolutions per second, it will be emitting radiation at that frequency *f* because its electric field, as seen from some fixed point a meter or so away, say, will be rotating *f* times per second.

On the other hand, the angular momentum quantization

condition must be true for all circular orbits of the electron around the proton, even for this very large atom. Furthermore, the radiation emitted must still be given by the difference in energies of neighboring orbits,

$$hf = E_{n+1} - E_n.$$

But $E_{n+1} - E_n$, the energy *spacing* between neighboring orbits, *depends on K.*

Therefore, Bohr concluded *K is fixed* by requiring that the frequency of radiation emitted by a really large atom be correctly given by ordinary common sense-that is, the frequency of the radiation must be the same as the orbital frequency of the electron, the number of cycles a second. In other words, for a large orbit we must have

$$E_{n+1} - E_n = hf = h.v/2\pi r$$

Here v is the speed of the electron in the orbit, and the orbit radius is r.

The strategy is then as follows: We assume that the only allowed orbits are those having angular momentum integral multiples of Kh, where K is some constant, so the nth orbit has angular momentum nKh. We can then use the equation of motion to determine the radius r_n, the electron speed v_n and the energy E_n for the nth orbit. Naturally, these all depend on K.

Now concentrate on orbits close to one meter in radius. They will radiate at the frequency given by the equations of motion for an electron circling a proton one meter away.

But this must *match up* with the frequency given by the energy difference between neighboring orbits divided by h. Now, if K is very small, this energy difference is small, and they won't match. If K is very large they won't match either. We must find the value of K for which these frequencies do match. That is what we do below in detail. We establish below that there can only be agreement between the classical radiation frequency for a man-sized atom and Bohr's prediction if $K = 1/2\pi$. Therefore, we must assume the angular momentum is always quantized in chunks of size $h/2\pi$.

It follows that the allowed energy levels are

$$E_n = -\frac{1}{4\pi\varepsilon_0}.\frac{e^2}{2r_n} = \left(\frac{1}{4\pi\varepsilon_0}\right)^2.\frac{me^4}{2K^2h^2}.\frac{1}{n^2}$$

$$= -\left(\frac{1}{4\pi\varepsilon_0}\right)^2 . \frac{2\pi^2 me^4}{h^2} . \frac{1}{n^2} .$$

Putting this together with $E_n - E_m = hf$ we get the Balmer formula:

$$f = \frac{E_n - E_2}{h} = \left(\frac{1}{4\pi\varepsilon_0}\right)^2 . \frac{2\pi^2 me^4}{h^3} . \left(\frac{1}{4} - \frac{1}{n^2}\right) .$$

The new point is that there is no adjustable parameter! The Rydberg constant that appeared before is here given in terms of h, m and e. The rather abstract argument that the quantum predictions must match the known classical results for large slow systems actually fixes the Rydberg constant. That is to say,

$$R_H = \left(\frac{1}{4\pi\varepsilon_0}\right)^2 . \frac{2\pi^2 me^4}{ch^3}$$

This formula was found to be correct within the limits of experimental error in measuring the quantities on the right. This matching for large systems is called the Correspondence Principle: in that limit, quantum predictions must correspond to known classical results.

Derivation of the Angular Momentum Quantization from the Correspondence Principle

Let us assume this large orbit is the nth (where n is of order 10^5!), so it has angular momentum

$$mvr = L = nKh.$$

Using

$$\frac{mv^2}{r} = \frac{1}{4\pi\varepsilon_0} . \frac{e^2}{r^2} ,$$

we find

$$m\sqrt{\frac{1}{4\pi\varepsilon_0} . \frac{e^2}{mr}} . r = nKh$$

from which we find the radii of the allowed orbits are given by

$$r_n = \frac{4\pi\varepsilon_0}{me^2} . n^2 K^2 h^2$$

Therefore the allowed energies are:

$$E_n = -\frac{1}{4\pi\varepsilon_0}.\frac{e^2}{2r_n} = -\left(\frac{1}{4\pi\varepsilon_0}\right).\frac{me^4}{2K^2h^2}.\frac{1}{n^2}$$

Thus for *n very large*:

$$E_{n+1} - E_n = -\left(\frac{1}{4\pi\varepsilon_0}\right)^2.\frac{me^4}{2K^2h^2}.\left(\frac{1}{(n+1)^2} - \frac{1}{n^2}\right)$$

$$\cong \left(\frac{1}{4\pi\varepsilon_0}\right)^2.\frac{me^4}{2K^2h^2}.\frac{2}{n^3}$$

Now this must be equal to $hf = hv/2\pi r$ for the appropriate v, r for this large orbit.

Using $mv_n r_n = nKh$, $v_n = nKh/mr_n$.

Thus $hv/2\pi\, r = nKh^2/2\pi\, mr_n^{\ 2}$.

Putting now

$$E_{n+1} - E_n = hf = h.v/2\text{p}\, r = nKh^2/2\text{p}\, mr_n^{\ 2}$$

And using

$$R_n = -\left(\frac{4\pi\varepsilon_0}{me^2}\right).n^2 2K^2h^2$$

gives

$$E_{n+1} - E_n \cong \left(\frac{1}{4\pi\varepsilon_0}\right)^2.\frac{me^4}{2K^2h^2}.\frac{2}{n^3}$$

$$= \frac{nKh^2}{2\pi mr_n^2} = \frac{nKh^2}{2\pi}.\left(\frac{1}{4\pi\varepsilon_0}\right)^2.\frac{me^4}{n^4K^4h^4}$$

Now, the second and fourth terms in the above equation must be equal in the limit of large n. Canceling out common factors of the two, we find that the condition for equality is:

$$K = 1/2\pi.$$

This was the argument Bohr used to establish that angular momentum for his model is quantized in units $h/2\pi$.

Chapter 2

Particles and Waves

FROM BOHR'S ATOM TO ELECTRON WAVES

Reactions to Bohr's Model

Bohr's interpretation of the Balmer formula in terms of quantized angular momentum was certainly impressive, but his atomic model didn't make much mechanical sense, as he himself conceded. For example, an electron jumping from the nth orbit to the mth emitted radiation at frequency equal to the energy difference of the orbits divided by h. Presumably, it began radiating as soon as it left its original orbit.

As Rutherford put it in a letter to Bohr, "how does an electron decide what frequency it is going to vibrate at when it passes from one stationary state to another? It seems to me that you would have to assume that the electron knows beforehand where it is going to stop."

A few days later, Bohr sent Rutherford another manuscript containing the correspondence principle argument gone. This was a bit too long for Rutherford. He responded: "As you know, it is the custom in England to put things very shortly and tersely in contrast to the Germanic method, where it appears to be a virtue to be as long-winded as possible". The Germans themselves were very skeptical of Bohr's model. Harald Bohr wrote to his brother from Gφttingen in the fall of 1913 that the young physicists there considered Bohr's model too "bold" and "fantastic".

Mysterious Spectral Lines

One good reason for the Germanic skepticism was the recent

discovery of apparently new spectral lines for hydrogen corresponding to *half*-integers in the Balmer formula. These new lines had been seen by a spectroscopist, Alfred Fowler, in a discharge tube containing a mixture of hydrogen and helium, and also in the spectra of a star by Pickering.

When he heard about these new lines, Bohr realized that his formula for the Rydberg constant,

$$R_H = \left(\frac{1}{4\pi\varepsilon_0}\right)^2 . \frac{2\pi^2 me^4}{ch^3}$$

should also apply to the ionized helium atom-that is, a helium nucleus with a single electron in orbit around it, provided e^2 is replaced by $2e^2$, to correct for the doubly charged nucleus. This gives an overall factor of 4, exactly equivalent to replacing the integers in the denominator by half-integers. In other words, Bohr argued, the supposed new hydrogen lines in fact were from ionized helium.

Fowler, however, was a very precise experimentalist. He could measure the spectral lines to five significant figures. Therefore, despite uncertainties in the values of e, m and h, the ratio of the "Rydberg constant" for helium to that for hydrogen could be measured to great accuracy. Fowler found it wasn't 4, but actually 4.0016.

Bohr's response to this news was conclusive. He pointed out that his analysis of the circular motion had neglected the finite mass of the nucleus. He should really have taken the electron to have an effective mass equal to $mM/(m + M)$. It is easy to verify that if this correction is made for both hydrogen and helium, the ratio changes from 4 to 4.0016!

This discovery won over most physicists to the point of view that Bohr was definitely on to something. Sir James Jeans remarked that the only justification for Bohr's postulates was "the very weighty one of success". According to Rosenfeld (Introduction to *Niels Bohr Collected Works*):

"At Gφttingen, that high place of mathematics and physics, where the sense for propriety was strong, the prevailing impression was one of scandal, or at least bewilderment, before the undeserved success of such high-handed disregard of the canons of formal logic." But all Germans didn't feel that way. Bohr's friend George Hevesy told Einstein about Bohr's identification of the Pickering-Fowler lines with helium. Einstein said: "This is an *enormous achievement*. The theory of Bohr must then be right."

A PERIODIC TABLE PUZZLE AND THE X-RAY CONNECTION

Mendeleev's success in demonstrating that the elements had recurring patterns of chemical behaviour when arranged in order of increasing atomic weight had some anomalies. In particular, cobalt has a higher atomic weight than nickel, yet its chemical properties strongly suggested it should come *before* nickel in the table.

It was gradually realized over the period from 1907 to 1913 (primarily by a Dutchman, van den Broek, actually a real estate lawyer who did physics in his spare time) that the significant parameter from a chemical point of view was not the mass number A, but the total number of electrons, and hence the charge on the nucleus Z, once the nuclear model was accepted.

The problem was measuring the nuclear charge. This could in fact be done by Rutherford scattering, but that was a long and tedious process. X-rays gave a quicker way. Any substance on being bombarded with sufficiently fast electrons emits a continuum of x-ray frequencies up to a maximum frequency f given by $hf =$ kinetic energy of electron, *plus some sharply defined lines*. The frequency of the sharply defined lines gradually increases with atomic number.

These x-ray lines were first discovered by Max von Laue in Munich in the summer of 1912. Harry Moseley, who worked with Rutherford mostly teaching undergraduates, decided to investigate this new spectroscopy. When Bohr came to visit in 1913, he encouraged Moseley to investigate x-ray spectra of many elements with a view to pinning down the values of Z. Moseley used a cathode ray tube a yard long with a little train inside it having a different element in each car, so many readings could be taken in one run.

He found several different x-ray lines for each element, with very similar patterns for each element, although gradually shifting to higher frequencies with increasing Z. For example, one particular line, labeled $K\alpha$, had a Z-dependent frequency:

$$f(Z) = R_H(Z-1)^2\left(\frac{1}{1^2}-\frac{1}{2^2}\right)$$

This made it possible to order all the elements unambiguously. It was also good evidence that Bohr's theory was close to correct for the innermost electrons in the atom, those in the field of the unshielded nucleus. Actually, the appearance of $Z - 1$ rather than Z in the formula is because the innermost shell has just two electrons, and they partially shield the nuclear charge from each other. The appearance of this factor was not understood at the time, because it was thought that there were more electrons in the innermost shell.

Bohr later remarked that Moseley's extremely clear x-ray work had far more impact than Rutherford's scattering experiment in widening acceptance of the Bohr-Rutherford atom. Tragically, Moseley died two years later, killed by a bullet in the head at Gallipoli.

What Bohr's Model Achieved

As Bohr himself emphasized, his model atom didn't *explain* anything, in the way classical dynamics or electromagnetism could explain how systems functioned in terms of a few underlying laws. What the Bohr model did do was to tie together previously unrelated phenomena, and therefore clarify the agenda for future investigations. It showed there was a link between the Balmer formula, Planck's constant and the nuclear atom, with quantized angular momentum playing a central role.

The precision of the spectral predictions for *ionized* helium demonstrated something was right about the model, as did the interpretation of characteristic x-rays for many elements in terms of inner-shell electrons.

On the other hand, despite heroic efforts over several years by Bohr and others, there was no successful prediction of the observed spectral lines for the next simplest atom neutral helium. Evidently, to make further progress in understanding atoms, some new insight was needed.

At this point the war intervened. Much of the momentum in physics was lost. Rutherford's lab was practically emptied, Rutherford himself spent most of his time on antisubmarine research. Bohr put much effort into establishing a physics institute in Copenhagen. He also thought at length about how the periodic table might be understood in terms of filling an atom with electrons.

There was something of a breakthrough in 1924 when the German physicist Wolfgang Pauli suggested there must be a law that two electrons couldn't go into the same orbit. (By this point, the number of allowed orbits had been substantially increased by Sommerfeld to include Keplerian elliptic ones.) Pauli's "exclusion principle" was introduced as an empirical rule.

We shall see later that it follows very naturally from symmetry considerations. We shall also find that in the full quantum picture of the atom, although the angular momentum is quantized, as Bohr realized, the orbital angular momentum is actually *zero* in the ground state of the hydrogen atom.

One of the more surprising results of the full quantum theory is that the energy levels found by Bohr, and even relativistic corrections computed by Sommerfeld, are exactly correct, even though the model of classical inverse square orbits is far from the truth (except for very large orbits).

A Student Prince Catches a Wave

Despite all the effort in Copenhagen and Gφttingen, the next real advance in understanding the atom came from an unlikely quarter a student prince in Paris. Prince Louis de Broglie was a member of an illustrious family, prominent in politics and the military since the 1600's.

Louis began his university studies with history, but his elder brother Maurice studied x-rays in his own laboratory, and Louis became interested in physics. He worked with the very new radio telegraphy during the war.

After the war, de Broglie focused his attention on Einstein's two major achievements, the theory of special relativity and the quantization of light waves. He wondered if there could be some connection between them. Perhaps the quantum of radiation really should be thought of as a particle. If so, the theory of special relativity suggested that the observed energy-momentum relationship, $E = cp$, meant that it had a very small rest mass, for it was always observednaturallyto be moving at the speed of light.

De Broglie suspected that if the speed of a sufficiently low energy quantum could be measured, it would be found to be less than c. On this point he was wrong (as far as we know!). Nevertheless, it was a very valuable conceptual breakthrough to think of the quantum of radiation as a *particle*, knowing full well that radiation is a wave.

In fact, his incorrect idea that the photon (as we now call the light quantum) had a rest mass led him to analyse the relationship between particle properties and wave properties by transforming to the rest frame of the photon, and he discovered that the energy and momentum of the particle were related to the frequency and wavelength of the wave by:

$$E = hf, p = h/\lambda$$

Of course, the first condition is the Planck-Einstein quantization, and the second follows trivially from it if we take $E = cp$ and $\lambda f = c$. But de Broglie showed it was more generally true it worked even if the photon had a rest mass.

Having decided that the photon might well be a particle with a rest mass, even if very small, it dawned on de Broglie that in other respects it might not be too different from other particles, especially the very light electron. In particular, maybe the electron also had an associated wave. The obvious objection was that if the electron was wavelike, why had no diffraction or interference effects been observed?

But there was an answer. If de Broglie's relation between momentum and wavelength, $p = h/\lambda$, also held for electrons, the wavelength was sufficiently short that these effects would be easy to miss. As de Broglie himself pointed out, the wave nature of light isn't very evident in everyday life, or in ray tracing in geometrical optics.

He suspected the apparently pure particle nature of electronic trajectories was analogous to the apparent straight line propagation of rays of light, over distance scales much greater than the wavelength.

However, the wavelike properties should be important on an atomic scale. No progress had been made in a decade in understanding why the electronic orbits in the Bohr atom were restricted to integral values of the angular momentum in units of h. But if the electron were in some sense a wave, it would be very natural to restrict the orbits to those of standing waves, for otherwise the electron wave on going around the orbit would interfere with itself destructively.

Suppose now the electron, having momentum p, is moving in a circular orbit of radius r. Then for a standing wave, a whole number of wavelengths must fit around the circle, so for some integer n, $n\lambda = 2\pi r$. Putting this together with $p = h/\lambda$, we find:

$$2\pi r = n\lambda = nh/p$$

so

$$L = pr = nh/2\pi.$$

The "standing wave" condition immediately gives Bohr's quantization of angular momentum!

This was the prince's Ph. D. thesis, presented in 1924. His thesis advisor was somewhat taken aback, and wasn't sure if this was sound work. He asked de Broglie for an extra copy of the thesis, which he sent to Einstein. Einstein wrote shortly afterwards: *"I believe it is a first feeble ray of light on this worst of our physics enigmas"*.

An Accident at the Phone Company Makes Everything Crystal Clear

There was an accident at the Bell Telephone Laboratories in April 1925. Clinton Davisson and L. H. Germer, looking for ways to improve vacuum tubes, were watching how electrons from an electron gun in a vacuum tube scattered off a flat nickel surface. Suddenly, while the experiment was running and the nickel target was very hot, a bottle of liquid air near the apparatus exploded, smashing one of the vacuum pipes, and air rushed into the apparatus.

The hot nickel target oxidized immediately. The layer of oxide made their target useless for further investigations. They decided to clean off the oxide by heating the nickel in a hydrogen atmosphere then in vacuum. After doing this for a prolonged period, the nickel looked good, and they resumed the investigation.

To their amazement, the pattern of electron scattering from the newly cleaned nickel target was completely different from that before the accident. What had changed? On examining their newly cleaned crystal carefully, they found a clue. The original target was polycrystalline made up of a multitude of tiny crystals, oriented randomly. During the prolonged heating of the cleaning process, the nickel had re-crystallized into a few large crystals.

To quote from their paper: "It seemed probable to us from these results that the intensity of scattering from a single crystal would exhibit a marked dependence on crystal direction, and we set about at once preparing experiments for an investigation of this dependence. We must admit that the results obtained in these experiments have proved to be quite at variance with our expectations.

It seemed likely that strong beams would be found issuing from the crystal along what may be termed its transparent directions the directions in which the atoms in the lattice are arranged along the smallest number of lines per unit area. Strong beams are indeed found issuing from the crystal, but only when the speed of bombardment lies near one or another of a series of critical values, ant then in directions quite unrelated to crystal transparency.

"The most striking characteristic of these beams is a one to one correspondence...which the strongest of them bear to the Laue beams that would be found issuing from the same crystal if the incident beam were a beam of x-rays. Certain others appear to be analogues... of optical diffraction beams from plane reflection gratings the lines of these gratings being lines or rows of atoms in the surface of the crystal.

Because of these similarities... a description... in terms of an equivalent wave radiation... is not only possible, but most simple and natural. This involves the association of a wavelength with the incident electron beam, and this wavelength turns out to be in acceptable agreement with the value *h*/*mv* of the undulatory mechanics, Planck's action constant divided by the momentum of the electron.

"That evidence for the wave nature of particle mechanics would be found in the reaction between a beam of electrons and a single crystal was predicted by Elsasser two years agoshortly after the appearance of L. de Broglie's original papers on wave mechanics."

WAVE PACKETS

THE WAVE-PARTICLE PUZZLE

De Broglie's explanation of the Bohr atom quantization rules, together with the accidental discovery of electron diffraction scattering by Davisson and Germer, make a very convincing case for the wave nature of the electron. Yet the electron certainly behaves like a particle sometimes.

An electron has a definite mass and charge, it can move slowly, it can travel through a piece of apparatus from a gun to a screen. What, then, is the relationship between the wave and particle viewpoints? De Broglie himself always felt both were always present. He called the wave a pilot wave, and thought it guided the motion of the particle. Unfortunately, that viewpoint leads to contradictions.

The standard modern interpretation is that the intensity of the wave (measured by the square of its amplitude) at any point gives the relative probability of finding the particle at that point. This interpretation, originally presented by Max Born in 1926, is parallel to the relation between the electromagnetic field and quanta the probability of finding a quantum (photon) at any point is proportional to the energy density of the field at that point, which is the square of the electric field vector plus the square of the magnetic field vector. The standard notation for the de Broglie wave function associated with the electron is y (x,t).

Thus, $|y\ (x,t)|^2 Dx$ is the relative probability of finding the electron in a small interval of length Dx near point x at time t. (For the moment, we restrict the electron to move in one dimension for simplicity. The generalization is straightforward.)

Keeping the Wave and the Particle Together?

Suppose following de Broglie we write down the relation between the "particle properties" of the electron and its "wave properties":

$$\tfrac{1}{2}mv^2 = E = hf,\ mv = p = h/\lambda$$

It would seem that we can immediately figure out the speed of the wave, just using l *f* = *cÁ,* say. We find:

$$l\,f = (h/mv).\ (\tfrac{1}{2}mv^2/h) = \tfrac{1}{2}v.$$

So the speed of the wave seems to be only half the speed of the electron!

How could they stay together? What's wrong with this calculation?

Localizing an Electron

To answer this question, it is necessary to think a little more carefully about the wave function corresponding to an electron traveling through a vacuum tube, say. The electron leaves the cathode, shoots through the vacuum, and impinges on an anode of a grid. At an intermediate point in this process, it is moving through the vacuum and the wave function must be nonzero over some volume, but zero in the places the electron has not possibly reached yet, and zero in the places it has definitely left.

However, if the electron has a precise energy, say fifty electron volts, it also has a precise momentum. This necessarily implies that the wave has a precise wavelength. But the only wave with a *precise* wavelength l has the form

$$\psi(x, t) = A \sin(kx - \omega t)$$

where $k = 2\pi/\lambda$, and $\omega = 2\pi f$. The problem is that this plane sine wave extends to infinity in both spatial directions, so cannot represent a particle whose wave function is non zero in a limited region of space.

Therefore, to represent a localized particle, we must superpose waves having different wavelengths. The principle is best illustrated by superposing two waves with slightly different wavelengths, and using the trigonometric addition formula:

$$\sin((k - \Delta k)x - (\omega - \Delta\omega)t) + \sin((k + \Delta k)x - (\omega + \Delta\omega)t)$$

$$= 2\sin(k - \omega k)\cos((\Delta k)x - (\Delta\omega)t)$$

This formula represents the phenomenon of beats between waves close in frequency. The first term, sin(*kx*-w *t*), oscillates at the average of the two frequencies. It is modulated by the slowly varying second term, which oscillates once over a spatial extent of order $\pi/\Delta k$. This is the distance over which waves initially in phase at the origin become completely out of phase. Of course, going a further distance of order $\pi/\Delta k$, the waves will become synchronized again.

That is, beating two close frequencies together breaks up the continuous wave into a series of packets, the beats. To describe a single electron moving through space, we need a single packet. This can be achieved by superposing waves having a continuous distribution of wavelengths, or wave numbers within of order Δk, say of k. In this case, the waves will be out of phase after a distance of order $\pi/\Delta k$, but since they have many different wavelengths, they will never get back in phase again. (This is all covered properly in the theory of Fourier transforms. Here we are just trying to make it plausible.)

The Uncertainty Principle

It should be evident from the above argument that to construct a wave packet representing an electron localized in a small region of space, the component waves must get out of phase rapidly. This means their wavelengths cannot be very close together. In fact, it is not difficult to give a semi-quantitative estimate of the spread in wavelength necessary, just from a consideration of the two beating waves.

A packet localized in a region of extent Δx can be constructed of waves having k's spread over a range Δk, where $\Delta x \sim \pi/\Delta k$.

Now, $k = 2\pi/l$, and $p = h/\lambda$, so $k = 2\pi\, p/h$.

Therefore, $\Delta k = 2\pi\, \Delta p/h$, and $\Delta x \sim \pi/\Delta k \sim h/\Delta p$ (dropping the factor of 2).

Thus:

$$\Delta x \Delta p \sim h$$

This is Heisenberg's Uncertainty Principle.

Phase Velocity and Group Velocity: Keeping the Wave and Particle Together

Establishing that an electron moving through space must be represented by a wave packet also resolves the paradox that the velocity of the waves seems to be different from the velocity of the electron. The point is that the electron waves, like water waves but unlike electromagnetic waves, have differing phase and group velocities.

To see this, consider again the beating of two waves of slightly different wavelengths.

$$\sin((k-\Delta k)x-(\omega-\Delta\omega)t)+\sin((k+\Delta k)x-(\omega+\Delta\omega)t)=$$

$$2\sin(kx-\omega t)\cos((\Delta k)x-(\Delta\omega)t)$$

The waves described by the term $\sin(kx - \omega t)$ have velocity $\frac{1}{2}v$, as previously derived. But the envelope, the shape of the wave packet, has velocity $\Delta\omega/\Delta k$ rather than ω/k. These velocities would be the same if ω were linear in k, as it is for ordinary electromagnetic waves. But the $\omega - k$ relationship follows from the energy-momentum relationship for the (non-relativistic) electron, $E = \frac{1}{2}mv^2 = p^2/2m$.

So $dE/dp = p/m = v$. But $E = hf = h\omega/2p$, and $p = h/\lambda = hk/2\pi$.

Therefore, $\Delta\omega/\Delta k = dE/dp = v$. So the packet travels at the speed we know the electron must travel at, even though the wave peaks within the packet travel at one-half the speed.

PROBABILITIES, AMPLITUDES AND PROBABILITY AMPLITUDES

Here we review three kinds of double slit experiments *classical particles*, *classical waves*, and *quantum things*.

BULLETS

First we consider a double slit experiment with *bullets*. The source is a not very accurate machine gun, which sprays bullets towards two parallel slits, 1 and 2, in a sheet of armorplate, which may be covered one at a time with another piece of armor. The bullets are "detected" by plowing into a soft (but thick) wood screen some distance beyond the slits. We assume the bullets don't break up, you always find whole bullets in the wood.

If we now cover slit 2, and fire a lot of bullets, by examining the screen at the end we will see where the bullets went, and be able to construct a probability distribution $P_1(x)dx$ giving the probability that a bullet going through slit 1 lands in the interval

x, $x + dx$. Similarly, covering slit 1, we can find the probability distribution $P_2(x)$ for bullets going through slit 2. Now suppose we open both slits this gives a probability distribution $P_{12}(x)$ and it is clear that this will be simply the sum:

$$P_{12}(x) = P_1(x) + P_2(x).$$

This of course follows from the self-evident fact that any bullet getting to the screen must have gone either through slit 1 or slit 2. *The probabilities add.*

There is one other important point about bullets they are lumpy. You either find one or you don't. This is also true of photons and electrons.

WATER WAVES

Now think about a double slit experiment with waves, say, surface waves on water the standard ripple tank experiment. Assume that plane waves of definite wavelength impinge from the left on a barrier having two slits P_1 and P_2, with P_2 being initially closed. Then to the right of the barrier, waves emanate from the open slit with circular wavefronts. These waves are detected at a screen over to the right. The waves could be detected by having corks bobbing up and down in the water just in front of the screen.

We could at some instant in time measure the height distribution $h_1(x)$ of the corks from waves coming through slit 1 (slit 2 being closed). Of course, this height distribution is changing all the time, and is not really what interests us. What we really want to know is the intensity of the wave motion at each point on the screen (analogous to the brightness of light, for example). In any simple harmonic motion, the intensity is proportional to the square of the amplitude. For our surface waves on water, the amplitude is of course just the height of the wave (relative to still water), so the energy intensity $I_1(x)$ for waves coming through slit 1 has the form

$$I_1(x) = Ah_1^2(x).$$

What happens when both slits are open? It is well known that waves superpose linearly—that is, the *heights* add. If at some instant of time with both slits open the height distribution of the corks is given by $h_{12}(x)$, then

$$h_{12}(x) = h_1(x) + h_2(x).$$

In this harmonic motion, the corks bob downwards as much as upwards, so h_1 and h_2 are as likely to cancel as they are to reinforce each other. The energy intensity when both slits are open,

$$I_{12}(x = Ah_{12}^2(x) = I_1(x) + I_2(x) + Ah_1(x)h_2(x).$$

The crucial point here is that in contrast to the bullets, the intensity when both slits are open is *not* the sum of the separate intensities for the slits being open one at a time. This is just the familiar wave interference effect.

Notice that in this classical wave picture, there is no lumpiness. The energy of the wave lapping against the screen varies smoothly, and can have any value we choose.

CLASSICAL LIGHT

Historically, an experiment equivalent to the double slit experiment for light was what convinced people it was indeed a wave. The light reaching a screen exhibited the identical intensity pattern observed for water waves.

This was fully explained with the advent of Maxwell's equations: the light was simply an electromagnetic wave, the amplitude of the electric field being analogous to the height of the water wave in the discussion above. To find the total electric field E_{12} at any point on the screen, the electric field E_1 of the wave emanating from S_1 is added to E_2 from S_2. The intensity of light I_{12} reaching the screen is proportional to E_{12}^2. Just as for the water waves, E_1 and E_2 oscillate about mean values of zero, so the analogy is complete.

PHOTONS

The discovery that light is after all lumpy on a fine enough scale throws this picture into confusion. Experimentally, light of frequency *f* can only be detected in lumps of size *hf*.

To the classically trained mind, this naturally suggests that the source of light is actually sending out these lumps, and each lump presumably passes through one of the slits on its way to the screen. However, this naove picture leads to a contradiction—how could it generate the observed diffraction pattern? Experimentally, the diffraction pattern builds up even if the light is so dim that there is a time interval between each photon (lump) passing through the apparatus. The picture of each photon going through one of the slits cannot be correct, because the distance between the slits governs the brightness pattern on the screen, and thus the probability pattern of where the photons land.

The only known resolution to this conceptual problem is to give up trying to visualize the path of the photon, just solve Maxwell's equation for the wave propagation, then interpret the resulting wave intensity as giving the relative probability of detecting the photon at any point. This is how quantum mechanics is done. It does not correspond to a physical picture readily interpretable in terms of familiar concepts. However, it does accord well with what is observed in nature.

ELECTRONS

Electrons behave in the same way as photons; there is little to add to the description. The irony is, of course, that before quantum mechanics electrons were seen as bullets, light as waves, and it has turned out that they are *both* basic quantum particles, with their propagation governed by *wave* equations: but when they arrive, they act like *particles.*

THE UNCERTAINTY PRINCIPLE

WAVES ARE FUZZY

The *wave nature* of particles implies that we cannot know *both* position *and* momentum of a particle to an arbitrary degree of accuracy if Δx represents the uncertainty in our knowledge of position, and that of momentum, then

$$\Delta p \Delta x \sim h$$

where h is Planck's constant. In the real world, particles are three-dimensional and we should say

$$\Delta p_x \Delta x \sim h$$

with corresponding equations for the other two spatial directions. The fuzziness about position is related to that of momentum *in the same direction.*

Let's see how this works by trying to measure y-position and y-momentum very accurately. Suppose we have a source of electrons, say, an electron gun in a CRT (cathode ray tube, such as an old-fashioned monitor). The beam spreads out a bit, but if we interpose a sheet of metal with a slit of width w, then for particles that make it through the slit, we know y with an uncertainty $\Delta y = w$. Now, if the slit is a long way downstream from the electron gun source, we also know p_y very accurately as the electron reaches the slit, because to make it to the slit the electron's velocity would have to be aimed just right.

But does the measurement of the electron's y position in other words, having it go through the slit affect its y momentum? The answer is *yes*. If it didn't, then sending a stream of particles through the slit they would all hit very close to the same point on a screen placed further downstream. But we know from experiment that this is *not* what happens a single slit diffraction pattern builds up, of angular width $\theta \sim \lambda/w$, where the electron's de Broglie wavelength λ is given by $px \equiv h/\lambda$ (there is a negligible contribution to λ from the y-momentum). The consequent uncertainty in p_y is

$$\Delta p_y/p_x \sim \theta \sim \lambda/w$$

Putting in $p_x = h/\lambda$, we find immediately that

$$\Delta p_y \sim h/w$$

so the act of measuring the electron's y position has fuzzed out its y momentum by precisely the amount required by the uncertainty principle.

TRYING TO BEAT THE UNCERTAINTY PRINCIPLE

In order to understand the Uncertainty Principle better, let's try to see what goes wrong when we actually try to measure position and momentum more accurately than allowed.

For example, suppose we look at an electron through a microscope. What could we expect to see? Of course, you know that if we try to look at something *really* small through a microscope it gets blurry a small sharp object gets diffraction patterns around its edges, indicating that we are looking at something of size comparable to the wavelength of the light being used.

If we look at something much smaller than the wavelength of light like the electron we would expect a diffraction pattern of concentric rings with a circular blob in the middle The size of the pattern is of order the wavelength of the light, in fact from optics it can be shown to be ~ $\lambda f/d$ where d is the diameter of the object lens of the microscope, f the focal distance (the distance from the lens to the object). We shall take $f/d \sim 1$, as it usually is. So looking at an object the size of an electron should give a diffraction pattern centered on the location of the object. That would seem to pin down its position fairly precisely.

What about the *momentum* of the electron? Here a problem arises that doesn't matter for larger objects the light we see has, of course, bounced off the electron, and so the electron has some recoil momentum. That is, by bouncing light off the electron we have given it some momentum.

Can we say how much? To make it simple, suppose we have good eyes and only need to bounce one photon off the electron to see it. We know the initial momentum of the photon (because we know the direction of the light beam we're using to illuminate the electron) and we know that after bouncing off, the photon hits the object lens and goes through the microscope, but we don't know *where* the photon hit the object lens.

The whole point of a microscope is that all the light from a point, light that hits the object lens *in different places*, is all focused back to one spot, forming the image (apart from the blurriness mentioned above).

So if the light has wavelength λ, its constituent photons have momentum $\sim h/\lambda$, and from our ignorance of where the photon entered the microscope we are uncertain of its x-direction momentum by an amount $\sim h/\lambda$. Necessarily, then, we have the same uncertainty about the electron's x-direction momentum, since this was imparted by the photon bouncing off.

But now we have a problem. In our attempts to minimize the uncertainty in the electron's momentum, by only using one photon to detect it, we are not going to see much of the diffraction pattern discussed above such diffraction patterns are generated by *many* photons hitting the film, retina or whatever detecting equipment is being used. A single photon generates a single point (at best!). This point will most likely be within of order λ of the centre of the pattern, but this leaves us with an uncertainty in position of order λ.

Therefore, in attempting to observe the position and momentum of a single electron using a single photon, we find an uncertainty in position $\Delta x \sim \lambda$, and in momentum $\Delta px \sim h/\lambda$. These results are in accordance with Heisenberg's Uncertainty Principle $\Delta x.\Delta p_x \sim h$.

Of course, we could pin down the position much better if we used N photons instead of a single one. From statistical theory, it is known that the remaining uncertainty $\sim \lambda / \sqrt{N}$. But then N photons have bounced off the electron, so, since each is equally likely to have gone through any part of the object lens, the uncertainly in momentum of the electron as a consequence of these collisions goes up as $\sqrt{N}$. (The same as the average imbalance between heads and tails in a sequence of N coin flips.)

Noting that the uncertainty in the momentum of the electron arises because we don't know where the bounced-off photon passes through the object lens, it is tempting to think we could just use a smaller object lens, that would reduce Δp_x. Although this is correct, recall from above that we stated the size of the diffraction pattern was $\sim\lambda f/d$, where d is the diameter of the object lens and f its focal length. It is easy to see that the diffraction pattern, and consequently Δx, gets bigger by just the amount that Δp_x gets smaller!

WATCHING ELECTRONS IN THE DOUBLE SLIT EXPERIMENT

Suppose now that in the double slit experiment, we set out to detect which slit each electron goes through by shining a light just behind the screen and watching for reflected light from the electron immediately after it had passed through a slit. Following the discussion in Feynman's *Lectures in Physics,* Volume III, we shall now establish that if we can detect the electrons, we ruin the diffraction pattern!

$$\left(n+\frac{1}{2}\right)\lambda_{\text{elec}} = d \sin \theta.$$

Taking the distance between the two slits to be d, the dark lines in the diffraction pattern are at angles

If the light used to see which slit the electron goes through generates an uncertainty in the electron's y momentum Δp_y, in order not to destroy the diffraction pattern we must have

$$\Delta p_y / p < \lambda_{\text{elec}} / d$$

(the angular uncertainty in the electron's direction must not be enough to spread it from the diffraction pattern maxima into the minima). Here p is the electron's full momentum, $p = h/\lambda_{\text{elec}}$. Now, the uncertainty in the electron's y momentum, looking for it with a microscope, is $\Delta p_y \sim h/\lambda_{\text{light}}$.

Substituting these values in the inequality above we find the condition for the diffraction pattern to survive is

$$\lambda_{\text{light}} > d,$$

the wavelength of the light used to detect which slit the electron went through must be greater than the distance between the slits. Unfortunately, the light scattered from the electron then gives one point in a diffraction pattern of size the wavelength of the light used, so even if we see the flash this does not pin down the electron sufficiently to say which slit it went through. Heisenberg wins again.

HOW THE UNCERTAINTY PRINCIPLE DETERMINES THE SIZE OF EVERYTHING

It is interesting to see how the actual physical size of the hydrogen atom is determined by the wave nature of the electron, in effect, by the Uncertainty Principle. In the ground state of the hydrogen atom, the electron minimizes its total energy. For a classical atom, the energy would be minus infinity, assuming the nucleus is a point (and very large in any case) because the electron would sit right on top of the nucleus. However, this cannot happen in quantum mechanics.

Such a very localized electron would have a very large uncertainty in momentum in other words, the kinetic energy would be large. This is most clearly seen by imagining that the electron is going in a circular orbit of radius r with angular momentum $h/2\pi$. Then one wavelength of the electron's de Broglie wave just fits around the circle, $\lambda_{\text{elec}} = 2\pi r$. Clearly, as we shrink the circle's radius r, λ_{elec} goes down proportionately, and the electrons momentum

$$p = h/\lambda_{elec} = h/2\pi r$$

increases. Adding the electron's electrostatic potential energy we find the total energy for a circular orbit of radius r is:

$$E(r) = K.E. + P.E. = \frac{p^2}{2m} - \frac{e^2}{4\pi\varepsilon_0 r} = \frac{h^2}{8m\pi^2 r^2} - \frac{e^2}{4\pi\varepsilon_0 r}.$$

Notice that for very large r, the potential energy dominates, the kinetic energy is negligible, and shrinking the atom lowers the total energy. However, for small enough r, the (always positive) kinetic energy term wins, and the total energy *grows* as the atom shrinks. Evidently, then, there must be a value of r for which *the total energy is a minimum*. Visualizing a graph of the total energy given by the equation above as a function of r, at the minimum point the slope of $E(r)$ is zero, $dE(r)/dr = 0$.

That is,

$$-\frac{h^2}{4m\pi^2 r} + \frac{e^2}{4\pi\varepsilon_0} = 0$$

giving

$$r_{min} = \frac{\varepsilon_0 h^2}{\pi m e^2}.$$

The total energy for this radius is the exact right answer, which is reassuring. The point of this exercise is to see that in quantum mechanics, unlike classical mechanics, a particle cannot position itself at the exact minimum of potential energy, because that would require a very narrow wave packet and thus be expensive in kinetic energy. The ground state of a quantum particle in an attractive potential is a trade off between potential energy minimization and kinetic energy minimization. Thus the physical sizes of atoms, molecules and ultimately ourselves are determined by Planck's constant.

SCHRODINGER'S EQUATION

Photons and Electrons

We have seen that electrons and photons behave in a very similar fashion both exhibit diffraction effects, as in the double slit experiment, both have particle like or quantum behaviour. We can in fact give a complete analysis of photon behaviour we can figure out how the electromagnetic wave propagates, using Maxwell's equations, then find the probability that a photon is in a given small volume of space *dxdydz*, is proportional to $|E|^2 dxdydz$, the energy density.

On the other hand, our analysis of the electron's behaviour is incomplete we know that it must also be described by a wave function $\psi(x, y, z, t)$ analogous to E, such that $|\psi(x, y, z, t)|^2\, dxdydz$ gives the probability of finding the electron in a small volume *dxdydz* around the point (x, y, z) at the time t. However, we do not yet have the analog of Maxwell's equations to tell us how ψ varies in time and space. A plausible derivation of such an equation by examining how the Maxwell wave equation works for a single-particle (photon) wave, and constructing parallel equations for particles which, unlike photons, have nonzero rest mass.

Maxwell's Wave Equation

Let us examine what Maxwell's equations tell us about the motion of the simplest type of electromagnetic wave a monochromatic wave in empty space, with no currents or charges present. First, we briefly review the derivation of the wave equation from Maxwell's equations in empty space:

$$\operatorname{div} \vec{B} = 0$$

$$\operatorname{div} \vec{E} = 0$$

$$\operatorname{curl} \vec{E} = -\frac{\partial \vec{B}}{\partial t}$$

$$\operatorname{curl} \vec{B} = \frac{1}{c^2}\frac{\partial \vec{E}}{\partial t}$$

To derive the wave equation, we take the curl of the third equation:

$$\operatorname{curl} \vec{E} = -\frac{\partial \vec{B}}{\partial t} \operatorname{curl} \vec{B} = -\frac{1}{c^2}\frac{\partial \vec{E}}{\partial t}$$

together with the vector operator identity

$$\text{curl curl} = \text{grad(div)} \times \nabla^2$$

to give

$$\nabla^2 \vec{E} - \frac{1}{c^2}\frac{\nabla^2 \vec{E}}{\partial t^2} = 0.$$

For a plane wave moving in the *x*-direction this reduces to

$$\frac{\nabla^2 \vec{E}}{\partial x^2} - \frac{1}{c^2}\frac{\nabla^2 \vec{E}}{\partial t^2} = 0$$

The monochromatic solution to this wave equation has the form

$$\vec{E}(x,t) = \vec{E}_0 e^{i(kx-\omega t)}.$$

(Another possible solution is proportional to $\cos(kx - \omega t)$. We shall find that the exponential form, although a complex number, proves more convenient. The physical electric field can be taken to be the real part of the exponential for the classical case.)

Applying the wave equation differential operator to our plane wave solution

$$\left(\frac{\partial^2}{\partial x^2} - \frac{1}{c^2}\frac{\partial^2}{\partial t^2}\right)\vec{E}_0 e^{i(kx-\omega t)} = \left(k^2 - \frac{\omega^2}{c^2}\right)\vec{E}_0 e^{i(kx-\omega t)} = 0$$

If the plane wave is a solution to the wave equation, this must be true for all *x* and *t*, so we must have.

$$\omega = ck.$$

This is just the familiar statement that the wave must travel at *c*.

What does the Wave Equation tell us about the *Photon*?

We know from the photoelectric effect and Compton scattering that the photon energy and momentum are related to the frequency and wavelength of the light by

$$E = h\nu = \hbar\omega$$

$$p = \frac{h}{\lambda} = \hbar k$$

Notice, then, that the wave equation tells us that $\omega = ck$ and hence $E = cp$. To put it another way, if we think of $e^{i(kx-\omega t)}$ as describing a particle (photon) it would be more natural to write the plane wave as

$$\vec{E}_0 e^{\frac{i}{\hbar}(px-Bt)}$$

that is, in terms of the energy and momentum of the particle.

In these terms, applying the (Maxwell) wave equation operator to the plane wave yields

$$\left(\frac{\partial^2}{\partial x^2} - \frac{1}{c^2}\frac{\partial^2}{\partial t^2}\right)\vec{E}_0 e^{\frac{i}{\hbar}(px-Bt)} = \left(p^2 - \frac{E^2}{c^2}\right)\vec{E}_0 e^{\frac{i}{\hbar}(px-Bt)} = 0$$

or

$$E^2 = c^2 p^2.$$

The wave equation operator applied to the plane wave describing the particle propagation yields the energy-momentum relationship for the particle.

Constructing a Wave Equation for a Particle with Mass

The discussion above suggests how we might extend the wave equation operator from the photon case (zero rest mass) to a particle having rest mass m_0. We need a wave equation operator that, when it operates on a plane wave, yields

$$E^2 = c^2p^2 + m_0^2c^4$$

Writing the plane wave function

$$\varphi(x,t) = Ae^{\frac{i}{\hbar}(px-Bt)}$$

where A is a constant, we find we can get $E^2 = c^2p^2 + m_0^2c^4$ by adding a constant (mass) term to the differentiation terms in the wave operator:

$$\left(\frac{\partial^2}{\partial x^2} - \frac{1}{c^2}\frac{\partial^2}{\partial t^2} - \frac{m_0^2c^2}{\hbar^2}\right)Ae^{\frac{i}{\hbar}(px-Bt)} =$$

$$-\frac{1}{\hbar^2}\left(p^2 - \frac{1}{c^2}\frac{E^2}{c^2} + m_0^2c^2\right)Ae^{\frac{i}{\hbar}(px-Bt)} = 0.$$

This wave equation is called the *Klein-Gordon* equation and correctly describes the propagation of relativistic particles of mass m_0. However, it's a bit inconvenient for nonrelativistic particles, like the electron in the hydrogen atom, just as $E^2 = m_0{}^2c^4 + c^2p^2$ is less useful than $E= p^2/2m$ for this case.

A Nonrelativistic Wave Equation

Continuing along the same lines, let us assume that a nonrelativistic electron in free space (no potentials, so no forces) is described by a plane wave:

$$\psi(x, t) = Ae^{\frac{i}{\hbar}(px-Bt)}.$$

We need to construct a wave equation operator which, applied to this wave function, just gives us the ordinary nonrelativistic energy-momentum relationship, $E = p^2/2m$. The p^2 obviously comes as usual from differentiating twice with respect to x, but the only way we can get E is by having a *single* differentiation with respect to time, so this looks different from previous wave equations:

$$i\hbar\frac{\partial\psi(x,t)}{\partial t} = -\frac{\hbar^2}{2m}\frac{\partial^2\psi(x,t)}{\partial x^2}.$$

This isSchrodinger *'s equation* for a free particle. It is easy to check that if $\psi(x, t)$ has the plane wave form given above, the condition for it to be a solution of this wave equation is just $E = p^2/2m$. Notice one remarkable feature of the above equation the i on the left means that ψ *cannot* be a real function.

How Does a Varying Potential Affect a de Broglie Wave?

The effect of a potential on a de Broglie wave was considered by Sommerfeld in an attempt to generalize the rather restrictive conditions in Bohr's model of the atom. Since the electron was orbiting in an inverse square force, just like the planets around the sun, Sommerfeld couldn't understand why Bohr's atom had only circular orbits, no Kepler-like ellipses.

De Broglie's analysis of the allowed circular orbits can be formulated by assuming that at some instant in time the spatial variation of the wave function on going around the orbit includes a phase term of the form $e^{\frac{i}{\hbar}pq}$, where here the parameter q measures distance around the orbit. Now for an acceptable wave function, the total phase change on going around the orbit must be $2n\pi$, where n is an integer. For the usual Bohr circular orbit, p is constant on going around, q changes by $2\pi r$, where r is the radius of the orbit, giving

$$\frac{1}{\hbar}p2\pi r = 2n\pi \ \psi$$

so

$$pr = n\hbar,$$

the usual angular momentum quantization.

What Sommerfeld did was to consider a general Kepler ellipse orbit, and visualize the wave going around such an orbit. Assuming the usual relationship $p = h/\lambda$, the wavelength will vary as the particle moves around the orbit, being shortest where the particle moves fastest, at its closest approach to the nucleus. Nevertheless, the phase change on moving a short distance Δq should still be $\frac{1}{\hbar}p\Delta q$, and requiring the wave function to link up smoothly on going once around the orbit gives

$$\int pdq = nh$$

Thus only certain elliptical orbits are allowed. The mathematics is nontrivial, but it turns out that every allowed elliptical orbit has the same energy as one of the allowed circular orbits. This is why Bohr's theory gave all the energy levels. Actually, this whole analysis is old fashioned (it's called the "old quantum theory") but we've gone over it to introduce the idea of *a wave with variable wavelength, changing with the momentum as the particle moves through a varying potential.*

SCHRODINGER'S EQUATION FOR A PARTICLE IN A POTENTIAL

Let us consider first the one-dimensional situation of a particle going in the x-direction subject to a "roller coaster" potential. What do we expect the wave function to look like? We would expect the wavelength to be shortest where the potential is lowest, in the valleys, because that's where the particle is going fastest—maximum momentum.

Perhaps slightly less obvious is that the amplitude of the wave would be largest at the tops of the hills (provided the particle has enough energy to get there) because that's where the particle is moving slowest, and therefore is most likely to be found.

With a nonzero potential present, the energy-momentum relationship for the particle becomes the energy equation

$$E = \frac{p^2}{2m} + V(x).$$

We need to construct a wave equation which leads naturally to this relationship. In contrast to the free particle cases discussed above, the relevant wave function here will no longer be a plane wave, since the wavelength varies with the potential. However, at a given x, the momentum is determined by the "local wavelength", that is,

$$p = -i\hbar \frac{\partial \psi}{\partial x}.$$

It follows that the appropriate wave equation is:

$$i\hbar \frac{\partial \psi(x,t)}{\partial t} = \frac{\hbar^2}{2m} \frac{\partial^2 \psi(x,t)}{\partial x^2} + V(x)\psi(x,t).$$

This is the standard one-dimensionalSchrodinger equation. In three dimensions, the argument is precisely analogous. The only difference is that the square of the momentum is now a sum of three squared components, for the x, y and z directions, so $\frac{\partial^2}{\partial x^2}$ becomes $\frac{\partial^2}{\partial x^2} + \frac{\partial^2}{\partial y^2} + \frac{\partial^2}{\partial z^2} = \nabla^2$, and the equation is:

$$i\hbar \frac{\partial \psi(x,y,z,t)}{\partial t} = \frac{\hbar^2}{2m} \nabla^2 \psi(x,y,z,t) + V(x,y,z)\psi(x,y,z,t).$$

This is the completeSchrodinger equation.

ELECTRON IN A BOX

PLANE WAVE SOLUTIONS

The best way to gain understanding ofSchrodinger 's equation is to solve it for various potentials. The simplest is a one-dimensional "particle in a box" problem. The appropriate potential is $V(x) = 0$ for x between 0, L and $V(x)$ = infinity otherwise—that is to say, there are infinitely high walls at $x = 0$ and $x = L$, and the particle is trapped between them.

This turns out to be quite a good approximation for electrons in a long molecule, and the three-dimensional version is a reasonable picture for electrons in metals. Between $x = 0$ and $x = L$ we have $V = 0$, so the wave equation is just

$$i\hbar\frac{\partial\psi(x,t)}{\partial t} = -\frac{\hbar^2}{2m}\frac{\partial^2\psi(x,t)}{\partial x^2}$$

A possible plane wave solution is

$$\psi(x, t) = Ae^{\frac{i}{\hbar}(px-Bt)}.$$

On inserting this into the zero-potentialSchrodinger equation above we find $E = p^2/2m$, as we expect. It is very important to notice that the complex conjugate, proportional to $e^{-\frac{i}{\hbar}(px-Bt)}$, is *not* a solution to theSchrodinger equation! If we blindly put it into the equation we get

$$E = -p^2/2m,$$

an unphysical result.

However, a wave function proportional to $e^{-\frac{i}{\hbar}(px+Bt)}$ gives $E = p^2/2m$, so this plane wave *is* a solution to the equation.

Therefore, the two allowed plane-wave solutions to the zero-potentialSchrodinger equation are proportional to $e^{-\frac{i}{\hbar}(px-Bt)}$ and $e^{-\frac{i}{\hbar}(px+Bt)}$ respectively.

Note that these two solutions have the *same time dependence* $e^{-\frac{iBt}{\hbar}}$.

To decide on the appropriate solution for our problem of an electron in a box, of course we have to bring in the walls—what they mean is that $\psi = 0$ for $x < 0$ and for $x > L$ because remember $|\psi|^2$ tells us the probability of finding the particle anywhere, and, since it's in the box, it's trapped *between* the walls, so there's zero probability of finding it outside. The condition $\psi = 0$ at $x = 0$

and $x = L$ reminds us of the vibrating string with two fixed ends—the solution of the string wave equation is standing waves of sine form. In fact, taking the difference of the two permitted plane-wave forms above gives a solution of this type:

$$\psi(x, \mathrm{t}) = A \sin \frac{px}{\hbar} e^{-\frac{iBt}{\hbar}}.$$

This wave function satisfies theSchrodinger equation between the walls, it vanishes at the $x = 0$ wall, it will also vanish at $x = L$ provided that the momentum variable satisfies:

$$\frac{pL}{\hbar} = \pi, 2\pi, 3\pi...$$

Thus the allowed values of p are $hn/2L$, where $n = 1, 2, 3...$, and from $E = p^2/2m$ the allowed energy levels of the particle are:

$$E = \frac{p^2}{2m} = \frac{1}{2m}\left(\frac{h}{2L}\right)^2, \frac{4}{2m}\left(\frac{h}{2L}\right)^2, \frac{9}{2m}\left(\frac{h}{2L}\right)^2,...$$

Note that these energy levels become more and more widely spaced out at high energies, in contrast to the hydrogen atom potential. (The harmonic oscillator potential gives equally spaced energy levels, so by studying how the spacing of energy levels varies with energy, we can learn something about the shape of the potential.) What about the overall multiplicative constant A in the wave function? This can be real or complex. To find its value, note that at a fixed time, say $t = 0$, the probability of the electron being between x and $x + dx$ is $|\psi|^2 dx$ or

$$|A|^2 \sin^2 \frac{px}{\hbar} dx.$$

The *total* probability of the particle being *somewhere* between 0, L must be unity:

$$\int_{x-0}^{x-L} |A|^2 \sin^2 \frac{px}{\hbar} dx = 1, \quad \text{so} \frac{1}{2} L |A|^2 = 1.$$

Hence

$$\psi(x, t) = \sqrt{\frac{2}{L}} \sin \frac{px}{\hbar} e^{-\frac{iBt}{\hbar}}.$$

When A is fixed in this way, by demanding that the total probability of finding the particle somewhere be unity, it is called the normalization constant.

STATIONARY STATES

Notice that at a later time the probability distribution for the wave function

$$\psi(x, t) = \sqrt{\frac{2}{L}} \sin\frac{px}{\hbar} e^{-\frac{iBt}{\hbar}}.$$

is the same, because time only appears as a phase factor in this time-dependent function, and so does not affect $|\psi|^2$.

A state with a time-independent probability distribution is called a stationary state.

States with Moving Probability Distributions

Recall that the Schrodinger equation is a linear equation, and the sum of any two solutions is also a solution to the equation. That means that we can add two solutions having different energies, and still have a legal wave function. We shall establish that in this case, the probability distribution varies in time. The simplest way to see how this must be is to look at an example. Let's add the ground state to the first excited state, and normalize the sum:

$$\psi(x, t) = \sqrt{\frac{1}{L}}\left(\sin\frac{\pi x}{L} e^{-\frac{izkt}{4mL^2}} + \sin\frac{2\pi x}{L} e^{-\frac{izkt}{mL^2}} \right)$$

note: h, not $\hbar$.

(You can check the normalization constant at $t = 0$). For general x, the two terms in the bracket rotate in the complex plane at different rates, so their sum has a time-varying magnitude. That is to say, $|\psi(x,t)|^2$ varies in time, so the particle must be moving around this is *not* a stationary state.

Exercise: To see this, note that at $t = 0$ the wave function is:

$$\psi(x, 0) = \sqrt{\frac{1}{L}}\left(\sin\frac{\pi x}{L} + \sin\frac{2\pi x}{L} \right)$$

and sketch this function: the particle is more likely to be found in the left-hand half of the box.

Now, suppose the time is $t = 4mL^2/h$, so $e^{\frac{izkt}{4mL^2}} = -1$. At this time,

$$\psi(4mL^2/h) = \sqrt{\frac{1}{L}}\left(-\sin\frac{\pi x}{L} + \sin\frac{2\pi x}{L}\right)$$

and it's easy to see that the particle is more likely to be found n the *right*-hand half.

That is to say, this wave function, a linear sum of wave functions corresponding to different energies, has a probability distribution that sloshes back and forth in the box: and, any attempt to describe a classical-type particle motion, bouncing back and forth, *necessarily* involves adding quantum wave functions of different energies.

Note that the frequency of the sloshing motion depends on the *difference* of the two energies: how constructively the two components interfere depends on the difference of the phases in the energies at the time. A single energy wave function always has a static probability distribution.

Of course, the *total* probability of finding the particle *somewhere* in the box remains unity: the normalization constant is time-independent.

THE TIME-INDEPENDENTSCHRODINGER EQUATION: EIGENSTATES AND EIGENVALUES

The only way to prevent $|\psi(x,t)|^2$ varying in time is to have all its parts changing phase in time at the same rate. *This means they all correspond to the same energy*. If we restrict our considerations to such *stationary states*, the wave function can be factorized

$$\psi(x, t) = \psi(x)\, e^{\frac{-iBt}{\hbar}}$$

and putting this wave function into theSchrodinger equation we find

$$-\frac{\hbar^2}{2m}\frac{d^2\psi(x)}{dx^2} + V(x)\psi(x) = E\psi(x).$$

This is the time-independentSchrodinger equation, and its solutions are the spatial wave functions for stationary states, states of definite energy. These are often called eigenstates of the equation. The values of energy corresponding to these eigenstates are called the eigenvalues.

AN IMPORTANT POINT: WHAT, EXACTLY, HAPPENS AT THE WALL?

Consider again the wavefunction for the lowest energy state of a particle confined between walls at $x = 0$ and $x = L$. The reader should sketch the wavefunction from some point to the left of $x = 0$ over to the right of $x = L$. To the left of $x = 0$, the wavefunction is exactly zero, then at $x = 0$ it takes off to the right (inside the box) as a sine curve. In other words, at the origin the slope of the wavefunction ψ is zero to the left, nonzero to the right. There is a discontinuity in the slope at the origin: this means the second derivative of ψ is infinite at the origin. On examining the time-independent Schrodinger equation above, we see the equation can *only* be satisfied at the origin because the potential becomes infinite there—the wall is an infinite potential. (And, in fact, since ψ becomes zero on approaching the origin from inside the box, the limit must be treated carefully.)

It now becomes obvious that if the box does not have infinite walls, but merely high ones, ψ describing a confined particle cannot suddenly go to zero at the walls: the second derivative must remain finite. For non-infinite walls, ψ and its derivative must be continuous on entering the wall.

This has the important physical consequence that ψ will be nonzero at least for some distance into the wall, even if classically the confined particle does not have enough energy to "climb the wall". (Which it doesn't, if it's confined.) Thus, in quantum mechanics, there is a non-vanishing probability of finding the particle in a region which is "classically forbidden" in the sense that it doesn't have enough energy to get there.

RUTHERFORD SCATTERING

RUTHERFORD AS ALPHA-MALE

In 1908 Rutherford was awarded the Nobel Prize - for

chemistry! The award citation read: "for his investigations into the disintegration of the elements, and the chemistry of radioactive substances."

While at McGill University, he had discovered that the radioactive element thorium emitted a gas which was itself radioactive, but if the gas radioactivity was monitored separately from the thorium's, he found it decreased geometrically, losing approximately half its current strength for each minute that passed. The gas he had found was a short-lived isotope of radon, and this was the first determination of a "half-life" for a radioactive material.

The chemists were of course impressed that Rutherford was fulfilling their ancient alchemical dream of transmuting elements, or at least demonstrating that it happened. Rutherford himself remarked at the ceremony that he "had dealt with many different transformations with various time-periods, but the quickest he had met was his own transformation from a physicist to a chemist". Still, Nobel prizes of any kind are nice to get, so he played along, titling his official Nobel lecture: "The chemical nature of the alpha-particle from radioactive substances". (He established that his favourite particle was an ionized helium atom by collecting alphas in an evacuated container, where they picked up electrons. After compressing this very rarefied gas, he passed an electric discharge through it and observed the characteristic helium spectrum in the light emitted.)

Rutherford was the world leader in alpha-particle physics. In 1906, at McGill, he had been the first to detect slight deflections of alphas on passage through matter. In 1907, he became a professor at the University of Manchester, where he worked with Hans Geiger. This was just a year after Rutherford's old boss, J. J. Thomson, had written a paper on his plum pudding atomic model suggesting that the number of electrons in an atom was about the same as the atomic number. (Not long before, people had speculated that atoms might contain thousands of electrons.

They were assuming that the electrons contributed a good fraction of the atom's mass.) The actual distribution of the electrons in the atom, though, was as mysterious as ever. Mayer's magnets were fascinating, but had not led to any quantitative conclusions on electronic distributions in atoms. Rutherford's 1906 discovery that his pet particles were slightly deflected on passing through atoms came about when he was finding their charge to mass ratio, by measuring the deflection in a magnetic field.

He detected the alphas by letting them impact photographic film. When he had them pass through a thin sheet of mica before hitting the film (so the film didn't have to be in the vacuum?) he found the image was blurred at the edges, evidently the mica was deflecting the alphas through a degree or two.

He also knew that the alphas wouldn't be deflected a detectable amount by the *electrons* in the atom, since the alphas weighed 8,000 times as much as the electrons, atoms contained only a few dozen electrons, and the alphas were very fast. The mass of the atom must be tied up somehow with the positive charge. Therefore, he reasoned, analyzing these small deflections might give some clue as to the distribution of positive charge and mass in the atom, and therefore give some insight into his old boss J. J.'s plum pudding. The electric fields necessary in the atom for the observed scattering already seemed surprisingly high to Rutherford.

SCATTERING ALPHAS

Rutherford's alpha scattering experiments were the first experiments in which individual particles were systematically scattered and detected. This is now the standard operating procedure of particle physics. To minimize alpha loss by scattering from air molecules, the experiment was carried out in a fairly good vacuum, the metal box being evacuated through a tube T. The alphas came from a few milligrams of radium (to be precise, its decay product radon 222) at R in the figure below, from the original paper, which goes on: "By means of a diaphragm placed at D, a pencil of alpha particles was directed normally on to the scattering foil F. By rotating the microscope [M] the alpha particles scattered in different directions could be observed on the screen S." Actually, this was more difficult than it sounds.

A single alpha caused a slight fluorescence on the zinc sulphide screen S at the end of the microscope. This could only be reliably seen by dark-adapted eyes (after half an hour in complete darkness) and one person could only count the flashes accurately for one minute before needing a break, and counts above 90 per minute were too fast for reliability. The experiment accumulated data from hundreds of thousands of flashes.

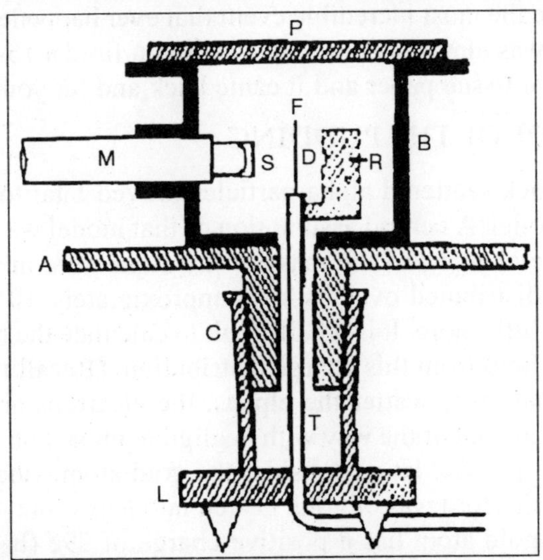

Rutherford's partner in the initial phase of this work was Hans Geiger, who later developed the Geiger counter to detect and count fast particles. Many hours of staring at the tiny zinc sulphide screen in the dark must have focused his mind on finding a better way!

In 1909, an undergraduate, Ernest Marsden, was being trained by Geiger. To quote Rutherford:

"I had observed the scattering of alpha-particles, and Dr. Geiger in my laboratory had examined it in detail. He found, in thin pieces of heavy metal, that the scattering was usually small, of the order of one degree. One day Geiger came to me and said, "Don't you think that young Marsden, whom I am training in radioactive methods, ought to begin a small research?"

Now I had thought that, too, so I said, "Why not let him see if any alpha-particles can be scattered through a large angle?" I may tell you in confidence that I did not believe that they would be, since we knew the alpha-particle was a very fast, massive particle with a great deal of energy, and you could show that if the scattering was due to the accumulated effect of a number of small scatterings, the chance of an alpha-particle's being scattered backward was very small. Then I remember two or three days later Geiger coming to me in great excitement and saying "We have been able to get some of the alpha-particles coming backward ..." It was quite the most incredible event that ever happened to me in my life. It was almost as incredible as if you fired a 15-inch shell at a piece of tissue paper and it came back and hit you."

DISPROOF OF THE PUDDING

The back scattered alpha-particles proved fatal to the plum pudding model. A central assumption of that model was that both the positive charge and the mass of the atom were more or less uniformly distributed over its size, approximately 10^{-10} meters across or a little more. It is not difficult to calculate the magnitude of electric field from this charge distribution. (Recall that this is the field that must scatter the alphas, the electrons are so light they will jump out of the way with negligible impact on an alpha.)

To be specific, let us consider the gold atom, since the foil used by Rutherford was of gold, beaten into leaf about 400 atoms thick. The gold atom has a positive charge of 79e (balanced of course by that of the 79 electrons in its normal state). Neglecting these electronsassume them scattered awaythe maximum electric force the alpha will encounter is that at the surface of the sphere of positive charge,

$$E.2e = \frac{1}{4\pi\varepsilon_0} \cdot \frac{79e.2e}{r_0^2} = 9.10^9 \cdot \frac{158.(1.6.10^{-19})^2}{10^{-20}}$$

$$= 3.64.10^{-6} \text{ newtons}$$

If the alpha particle initially has momentum p, for small deflections the angle of deflection (in radians) is given by (*delta_p*)/*p*, where *delta_p* is the sideways momentum resulting from the electrically repulsive force of the positive sphere of charge. Assuming the atomic sphere itself moves negligiblyit is much heavier than the alpha, so this is reasonablethe trajectory of the alpha in the inverse square electric field can be found by standard methods. It is the same mathematical problem as finding the elliptic orbits of planets around the sun. Replacing inverse square attraction with inverse square repulsion changes the orbit from an ellipse (or a hyperbola branch swinging around the sun for a comet) to a hyperbola branch lying on one side of the centre of repulsion.

In fact, one can get a clear idea of how much deflection comes about without going into the details of the trajectory. Outside the atom, the repulsive electrical force falls away as the inverse square. Inside the atom, the force drops to zero at the centre, just as the gravitational force is zero at the centre of the earth. The force is maximum right at the surface. Therefore, a good idea of the sideways deflection is given by assuming the alpha experiences that maximal force for a time interval equal to the time it takes the alpha to cross the atomsay, a distance $2r_0$.

Note that since the alpha particle has mass 6.7×10^{-27} kg, from $F = ma$, the electric force at the atomic surface above will give it a sideways acceleration of 5.4×10^{20} meters per sec per sec (compare $g = 10$!). But the force doesn't have long to act - the alpha is moving at 1.6×10^{7} meters per second. So the time available for the force to act is the time interval a particle needs to cross an atom if the particle gets from New York to Australia in one second.

The time $t_0 = 2r_0/v = 2\times10^{-10}/1.6\times10^{7} = 1.25\times10^{-17}$ seconds.

Thus the magnitude of the total sideways velocity picked up is the sideways acceleration multiplied by the time,

$1.25\times10^{-17}\times5.4\times10^{20} = 6750$ meters per second.

This is a few ten-thousandths of the alpha's forward speed, so there is only a very tiny deflection. Even if the alpha hit 400 atoms in succession and they all deflected it the same way, an astronomically improbable event, the deflection would only be of order a degree. Therefore, the observed deflection through *ninety* degrees and more was completely inexplicable using Thomson's pudding model!

EMERGENCE OF THE NUCLEUS

Rutherford pondered the problem for some months. He had been a believer in his former boss's pudding model, but he eventually decided there was simply no way it could generate the strength of electric field necessary to deflect the fast moving alphas. Yet it was difficult to credit there was much more positive charge around than that necessary to compensate for the electrons, and it was pretty well established that there were not more than a hundred or so electrons (we used 79, the correct value-that was not known exactly until a little later). The electric field from a sphere of charge reaches its maximum on the surface, as discussed above. Therefore, for a given charge, assumed spherically distributed, the only way to get a stronger field is to *compress it into a smaller sphere*. Rutherford concluded that he could only explain the large alpha deflections if the positive charge, and most of the mass of the atom, was in *a sphere* much smaller than the atom itself.

It is not difficult to estimate from the above discussion how small such a *nucleus* would have to be to give a substantial deflection. We found a sphere of radius 10^{-10} meters gave a deflection of about 4×10^{-4} radians. We need to increase this deflection by a factor of a few thousand. On decreasing the radius of the sphere of positive charge, the force at the surface increases as the inverse radius *squared*.

On the other hand, the *time* over which the alpha experiences the sideways force *decreases* as the radius. The *total deflection*, then, proportional to the product of force and time, *increases as the inverse of the radius*. This forces the conclusion that the positive charge is in a sphere of radius certainly less than 10^{-13} meters, provided all the observed scattering is caused by one encounter with a nucleus.

Rutherford decided that the observed scattering *was* in fact from a single nucleus. He argued as follows: since the foil is only 400 atoms thick, it is difficult to see how ninety degree scatterings could arise unless the scattering by a single nucleus was *at least* one degree, say 100 times that predicted by the Thomson model. This would imply that the nucleus had a radius at most one-hundredth that of the atom, and therefore presented a target area for one-degree scattering (or more) to the incoming alphas only one ten-thousandth that of the atom.

(In particle physics jargon, this target area is called the *scattering cross section*.) If an alpha goes through 400 layers of atoms, and in each layer it has a chance of one in ten thousand of getting close enough to the nucleus for a one-degree scatter, this is unlikely to happen twice. It follows that almost certainly only one scattering takes place. It *then* follows that all ninety or more degrees of scattering must be a single event, so the nucleus must be even smaller than one hundredth the radius of the atomit must be less than 10^{-13} meters, as stated above.

MODELING THE SCATTERING

To visualize the path of the alpha in such a scattering, Rutherford "had a model made, a heavy electromagnet suspended as a pendulum on thirty feet of wire that grazed the face of another electromagnet set on a table. With the two grazing faces matched in polarity and therefore repelling each other, the pendulum was deflected" into a hyperbolic path. In place of this rather substantial model, I've put in two applets, showing alphas deflected by two spheres of charge having the same total charge, but different radii.

The large sphere, then, represents the Thomson model, the smaller one a nuclear model. These are of course not to scale! An accurate nuclear representation would take far less than one pixel, and hours between significant scatterings. Nevertheless, the relation between large angle scattering and the size of the positive sphere is clear from the model.

SEEING THE NUCLEUS

Having decided that the observed scattering of the alphas came from single encounters with nuclei, and assuming that the scattering force was just the electrostatic repulsion, Rutherford realized that finding the scattering angle as a function of ingoing speed and *impact parameter* (how close to the centre would the alpha particle pass if the repulsion were switched off) was an exercise in Newtonian mechanics.

Although not exactly a hot shot theorist, Rutherford managed to figure this out after a few weeks. Anyway, the bottom line is that for a nucleus of charge Z, and incident alpha particles of mass m and speed v, the rate of scattering to a point on the screen corresponding to a scattering angle of *theta* (angle between incident velocity and final velocity of alpha) is proportional to:

$$\text{Scattering into small area} \propto \left(\frac{1}{4\pi\varepsilon_0}\cdot\frac{Ze^2}{mv^2}\right)^2 \frac{1}{\sin^4(\theta/2)}.$$

Analysis of the hundred thousand or more scattering events

recorded for the alphas on gold fully confirmed the angular dependence predicted by the above analysis. But it didn't work for aluminum! On replacing the gold foil by aluminum foil (some years later), it turned out that small angle scattering obeyed the above law, but large angle scattering didn't.

Rutherford correctly deduced that in the large angle scattering, which corresponded to closer approach to the nucleus, the alpha was actually hitting the nucleus. This meant that the size of the nucleus could be worked out by finding the maximum angle for which the inverse square scattering formula worked, and finding how close to the centre of the nucleus such an alpha came. Rutherford estimated the radius of the aluminum nucleus to be about 10^{-14} meters.

THE BEGINNINGS OF NUCLEAR PHYSICS

The First World War lasted from 1914 to 1918. Geiger and Marsden were both at the Western front, on opposite sides. Rutherford had a large water tank installed on the ground floor of the building in Manchester, to carry out research on defence against submarine attack. Nevertheless, occasional research on alpha scattering continued. Scattering from heavy nuclei was fully accounted for by the electrostatic repulsion, so Rutherford concentrated on light nuclei, including hydrogen and nitrogen. In 1919, Rutherford established that an alpha impinging on a nitrogen nucleus can cause a *hydrogen* atom to appear! Newspaper headlines blared that Rutherford had "split the atom".

Shortly after that experiment, Rutherford moved back to Cambridge to succeed J. J. Thomson as head of the Cavendish laboratory, working with one of his former students, Chadwick, who had spent the war years interned in Germany. They discovered many unusual effects with alpha scattering from light nuclei. In 1921, Chadwick and co-author Bieler wrote: "The present experiments do not seem to throw any light on the nature of the law of variation of the forces at the seat of an electric charge, but merely show that the forces are of great intensity ... It is our task to find some field of force which will reproduce these effects."

In fact, Rutherford was beginning to focus his attention on the actual construction of the nucleus and the alpha particle. He coined the word "proton" to describe the hydrogen nucleus, it first appeared in print in 1920. At first, he thought the alpha must be made up of four of these protons somehow bound together by having two electrons in the middle - this would get the mass and charge right, but of course nobody could construct a plausible electrostatic configuration.

Then he had the idea that maybe there was a special very tightly bound state of a proton and an electron, much smaller than an atom. By 1924, he and Chadwick were discussing how to detect this neutron. It wasn't going to be easy - it probably wouldn't leave much of a track in a cloud chamber. In fact, Chadwick did discover the neutron, but not until 1932, and it wasn't much like their imagined proton-electron bound state. But it did usher in the modern era in nuclear physics.

SCATTERING STATES AND BARRIERS

STREAMS OF PARTICLES

Our analysis of the time independentSchrodinger equation using the spreadsheet limited us to real values of the wave function $\psi(x)$. This is fine for analyzing bound states in a potential, or standing waves in general, but cannot be used, for example, to represent a stream of electrons being emitted by an electron gun, such as in an old TV tube. The reason is that a real wavefunction $\psi(x)$, in an energetically allowed region, is made up of terms locally like $\cos kx$ and $\sin kx$, multiplied in the full wave function by the time dependent phase factor $e^{-iBt/\hbar}$, giving equal amplitudes of right moving waves $e^{i(px-Bt)/\hbar}$ and left moving waves $e^{i(px+Bt)/\hbar}$. So if we are interested in a system in which there are not equal numbers of particles moving to the right and to the left, we must have a wave function such that even the x-dependent part is complex.

A simple example is a stream of particles of energy E moving from the left in one dimension through a region of zero potential, encountering an upward step potential V_0, where $V_0 < E$, at the origin $x = 0$, so that classically the particles would climb the hill and continue to the right. We shall represent the incoming wave function by a plane wave,

$$\psi(x, t) = Ae^{ikx}\, e^{-iBt/k} \text{ for } x < 0$$

It proves slightly more convenient to work with wave number k rather than particle momentum $p = \hbar k$ in scattering problems of this type. If we now think of the classical picture of a particle approaching a hill (smoothing off the corners a bit) that it definitely has enough energy to surmount, we would perhaps expect that the wave function continues beyond $x = 0$ in the form

$$\psi(x, t) = Be^{ik,x} e^{-iBt/k} \text{ for } x > 0,$$

where k_1 corresponds to the slower speed the particle will have after climbing the hill. Schrodinger's equation requires that the wave function have no discontinuities and no kinks (discontinuities in slope) so the $x < 0$ and $x > 0$ wave functions must match smoothly at the origin. For them to have the same value, we see from above that $A = B$. For them to have the same slope we must have $kA = k_1B$. Unfortunately, the only way to satisfy both these equations with our above wave functions is to take $k = k_1$ which means there is no step potential at all!

Question: what is wrong with the above reasoning? The answer is that we have been led astray by our mental picture of the particles as little balls rolling along in a potential, with enough energy to get up the hill, etc. Schrodinger 's equation is a *wave equation*. Building intuition about solutions should rely on experience with waves. We should be thinking about a light wave going from air into glass, for example. If we do, we realise that at *any* interface *some of the light gets reflected*. This means that our expression for the wave function for $x < 0$ is incomplete, we need to add a *reflected* wave, giving

$$\psi(x, t) = Ae^{ikx}e^{-iBt/k} + Ce^{-ikx}e^{-iBt/\hbar} \text{ for } x < 0$$

$$\psi(x, t) = Be^{ik_1x}e^{-iBt/k} \qquad \text{for } x \geq 0$$

If we now match the wave function and its derivative at the origin, we find

$$A + C = B$$

$$k(A - C) = k_1B.$$

Recalling that the square of the wave function denotes probability, it is easy to check that the fraction of the wave that is reflected

$$R = \frac{C^2}{A^2} = \left(\frac{k - k_1}{k + k_1}\right)^2.$$

Evidently, the fraction of the wave transmitted.

$$T = 1 - R = \frac{4kk_1}{(k + k_1)^2}.$$

Question: isn't the amount transmitted just given by B^2/A^2?

The answer is no. The ratio B^2/A^2 gives the relative probability of finding a particle in some small region in the transmitted stream relative to that in the incoming stream, but the particles in the transmitted stream are moving more slowly, by a factor k_1/k. This means that just comparing the densities of particles in the transmitted and incoming streams is not enough. The physically significant quantity is the *probability current* flowing past a given point, and this is the product of the density *and* the speed. Therefore, the transmission coefficient is B^2k_1/A^2k.

Exercise: prove that even a step *down* gives rise to some reflection.

BARRIERS

If a plane wave coming in from the left encounters a step at the origin of height $V_0 > E$, the incoming energy, there will be total reflection, but with an exponentially decaying wave penetrating some distance into the step. Suppose now we replace the step with a barrier,

$$V = 0 \text{ for } x < 0$$
$$V = V_0 \text{ for } 0 < x < L$$
$$V = 0 \text{ for } L < x.$$

In this situation, the wave function will still decay exponentially into the barrier (assuming the barrier is thick compared to the exponential decay length), but on reaching the far end at $x = L$, a plane wave solution is again allowed, so there is a nonzero probability of finding the particle beyond the barrier, moving with its original speed. This phenomenon is called *tunneling*, since in the classical picture the particle doesn't have enough energy to get over the top of the barrier.

ALPHA DECAY

A good example of tunneling, and one which helped establish the validity of quantum ideas at the nuclear level, is alpha decay.

Certain large unstable nuclei decay radioactively by emitting an alpha-particle, a tightly bound state of two protons and two neutrons. It is thought that alpha- particles may exist, at least as long lived resonances, inside the nucleus.

For such a particle, the strong but short ranged nuclear force creates a spherical finite depth well having a steep wall more or less coinciding with the surface of the nucleus. However, we must also include the electrostatic repulsion between the alpha-particle and the rest of the nucleus, a potential $(1/4\pi\varepsilon_0)\,(Z-2)\,2e^2/r$ outside the nucleus.

This means that, as seen from inside the nucleus, the wall at the surface may not be a step but a barrier, in the sense we used the word above, a step up followed by a slide down the electrostatic curve. Therefore, an alpha-particle bouncing around inside the nucleus may have enough energy to tunnel through to the outside world.

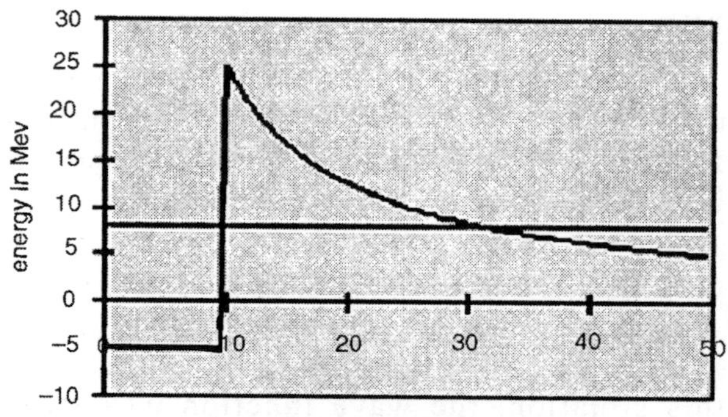

It is evident that the more energetic the alpha-particle is, the thinner the barrier it faces. Since the wave function decays exponentially in the barrier, this can make a huge difference in tunneling rates. It is not difficult to find the energy with which the alpha-particle hits the nuclear wall, because this will be the same energy with which it escapes.

Therefore, if we measure the energy of an emitted alpha, since we think we know the shape of the barrier pretty well, we should be able, at least numerically, to predict the tunneling rate. The only other thing we need to know is how many times per second alpha's bounce off the wall. The size of the nucleus is of order 10^{-14} meters, if we assume an alpha moves at, say, 10^7 meters per second, it will bang into the wall 10^{21} times per second. This is a bit handwaving, but all alpha-radioactive nuclei are pretty much the same size, so perhaps it's safe to assume this will be about the same for all of them.

If we do that, we get impressive agreement with experiment over a huge range of lifetimes. polonium^{212} emits alpha's with energy 8.95 MeV, and lasts 3 10^{-7} seconds, thorium^{232} emits 4.05 MeV alpha's, and lasts 1.4 10^{10} years. These can both be understood in terms of essentially the same barrier being tunneled through at the different heights corresponding to the alpha energy.

Chapter 3

Finite Depth Square Well

We have considered in some detail a particle trapped between infinitely high walls a distance L apart, we have found the wave function solutions of the time independentSchrodinger equation, and the corresponding energies. The essential point was that the wave function had to go to zero at the walls, because there is zero probability of finding the particle penetrating an infinitely high wall. This meant that the lowest energy state couldn't have zero energy, that would give a constant nonzero wave function.

Rather, the lowest energy state had to have the minimal amount of bending of the wave function necessary for it to be zero at the two walls but nonzero in between-this corresponds to half a period of a sine or cosine (depending on the choice of origin), these functions being the solutions ofSchrodinger 's equation in the zero potential region between the walls. The sequence of wave functions (eigenstates) as the energy increases have 0, 1, 2, ... zeros (nodes) in the well.

Let us now consider how this picture is changed if the potential at the walls is not infinite. It will turn out to be convenient to have the origin at the centre of the well, so we take

$$V(x) = V_0 \text{ for } x < -L/2$$
$$V(x) = 0 \text{ for } -L/2 < x < L/2$$
$$V(x) = V_0 \text{ for } L/2 < x.$$

Having the potential symmetric about the origin makes it easier to catalog the wave functions. For a symmetric potential, the wave functions can always be taken to be symmetric or antisymmetric. (If a wave function $\psi(x)$ is a solution ofSchrodinger 's equation with energy E, and the potential is symmetric, then $\psi(-x)$ is a solution with the same energy. This means that $\psi(x)+$

$\psi(-x)$ and $\psi(x) - \psi(-x)$ are also solutions, since the equation is linear, and these are symmetric and antisymmetric respectively, and using them is completely equivalent to using the original $\psi(x)$ and its reflection $\psi(-x)$.)

How is the lowest energy state wave function affected by having finite instead of infinite walls? Inside the well, the solution toSchrodinger 's equation is still of cosine form (it's a state symmetric about the origin). However, since the walls are now finite, $\psi(x)$ cannot change slope discontinuously to a flat line at the walls. It must instead connect smoothly with a function which is a solution toSchrodinger 's equation *inside the wall.*

The equation in the wall is

$$-\frac{\hbar^2}{2m}\frac{d^2\psi(x)}{dx^2} + V_0\psi(x) = E\psi(x)$$

and has two exponential solutions (say, for $x > L/2$) one increasing to the right, the other decreasing,

$$e^{\alpha x} \text{ and } e^{-\alpha x}, \text{ where } \alpha = \sqrt{2m(V_0 - E)/\hbar^2}.$$

(We are assuming here that $E < V_0$, so the particle is bound to the well. We shall find this is always true for the lowest energy state.)

Let us try to construct the wave function for the energy E corresponding to this lowest bound state.

From the equation with $V_0 = 0$, the wavefunction inside the well (let's assume it's symmetric for now) is proportional to $\cos kx$,

where $k = \sqrt{2mE)/\hbar^2}$.

The wave function (and its derivative!) inside the well must match a sum of exponential terms—the wave function in the *wall* at $x = L/2$, so

$$\cos(kL/2) = Ae^{\alpha L/2} + Be^{-\alpha L/2}$$
$$-k\sin(kL/2) = Ae^{\alpha L/2} - Be^{-\alpha L/2}.$$

(By writing just a cosine term inside the well, we have left out the overall normalization constant. This can be put back in at the end.)

Solving these equations for the coefficients A, B in the usual way, we find that in general the cosine solution inside the well goes smoothly into a linear combination of exponentially increasing and decreasing terms in the wall. (By the symmetry of the problem, the same thing must happen for $x < -L/2$.) *However, this cannot in general represent a bound state in the well.* The increasing solution increases *without limit* as x goes to infinity, so since the square of the wave function is proportional to the probability of finding the particle at any point, the particle is infinitely more likely to be found at infinity than anywhere else. It got away!

This clearly makes no sense we're trying to find wave functions for particles that stay in, or at least close to, the well. We are forced to conclude that the *only* exponential wave function that makes sense is the one for which *A is exactly zero*, so that there is only a *decreasing* wave in the wall.

Requiring the decreasing wave function, $A = 0$, means that only a discrete set of values of k, or E, satisfy the boundary condition equations above. They are most simply found by taking $A = 0$ and dividing one equation by the other to give:

$$\tan(kL/2) = \alpha/k.$$

This cannot be solved analytically, but is easy to solve graphically by plotting the two sides as functions of k (recall $\alpha = \sqrt{2m(V_0 - E)/\hbar^2}$, and $\sqrt{2mE)/\hbar^2}$) and finding where the curves intersect.

SOLVING SCHRODINGER 'S EQUATION NUMERICALLY

A more straightforward approach is to solve theSchrodinger equation numerically, using a spreadsheet, and explore how the wave function varies as E is changed. This method has the great advantage that the spreadsheet can be readily adapted to any reasonable potential. We present first the simplest, least sophisticated method, which gives an accuracy of around one percent for energy values using a few hundred rows of the spreadsheet.

To simplify adapting the spreadsheet to the equation, we choose units *so that* $\hbar^2/2m = 1$.

From now on, we shall denote the wave function by $f(x)$ or

just f instead of $\psi(x)$, to lessen headaches in setting up the spreadsheet.

Schrodinger's Equation is then:

$$\frac{d^2 f(x)}{dx^2} = (V(x) - E)\, f(x).$$

To solve this using the spreadsheet, it is necessary to discretize the derivative. We replace the continuous variable x by a sequence of equally spaced points dx apart, such as 0, dx, $2dx$, $3dx$, ... Here dx is of course finite, we should really call it Δx.

We approximate the second derivative by:

$$\left(\frac{d^2 f(x)}{dx^2}\right)_{x-x_j} = \frac{f(x_{j+1}) - 2f(x_j) + f(x_{j-1})}{(dx)^2}$$

so the differential equation is replaced by the difference equation:

$$\frac{f(x_{j+1}) - 2f(x_j) + f(x_{j-1})}{(dx)^2} = (V(x_j) - E)\, f(x_j).$$

Notice that from this equation, if we know $f(x_j)$, $f(x_{j-1})$ and $V(x_j)$ we can find $f(x_{j+1})$. Of course, we do know $V(x_j)$, this is the potential we are trying to solve for. We can also specify E, the energy of the particle, although we shall find that for negative E, only a discrete set of bound state energies correspond to normalizable wave functions.

A second-order differential equation like this needs *two* boundary conditions for the solution to be defined. We can specify the initial value of f and its derivative df/dx. But this is just equivalent to giving the first two members of the discrete series, $f(x_0)$ and $f(x_1)$, say. We can then use the difference equation to find $f(x_2)$, then use it again to find $f(x_3)$ and so on. This is what the spreadsheet does for us: we find $f(x_{j+1})$ row by row using the difference equation above written:

$$f(x_{j+1}) = (2 + (dx)^2\, (V(x_j) - E)) - f(x_{j-1})$$

USING THE SPREADSHEET FOR THE FINITE SQUARE WELL

Recall that by taking the origin in the centre of the square well, we argued that we need only look at wave functions that were symmetric or antisymmetric about the origin. If you try to sketch such wave functions, you will find that symmetric wave functions must have zero slope at the origin, and antisymmetric wave functions must be zero at the origin.

Furthermore, if we know the wave function in the right-hand half, that is, for $x > 0$, we know it for all x, from the symmetry. Hence, we need only solveSchrodinger 's equation going from the origin to the right - we can take $x_0 = 0$, $x_1 = dx$, etc.

On integrating out from the origin with energy E set less than V_0, once we cross over into the wall the solution is some combination of an exponentially increasing function and an exponentially decreasing function,

$$f(x) = A(E)e^{\alpha x} + B(E)e^{-\alpha x}$$

where the exponential coefficient α is positive, and depends on E it was defined in the preceding section.

Let us first look at the *symmetric* solutions for very low energies, so take $f(0) = 1, f'(0) = 0$. (Note that we cannot find the correct overall normalization constant until we find the solution, then integrate its square over all space this can always be done later, and is unnecessary for analyzing the properties of the state).

Let us begin with the trivial case $E = 0$. For zero energy, inside the well $E - V$ will of course be identically zero, so fromSchrodinger 's equation the *slope* of $f(x)$ can never change, consequently $f(x) = 1$ for all $x < L$. On reaching the wall, this wave function and its derivative connect smoothly from inside to outside if $A = B = ½$. It is clear that as we keep going to the right, the A term in the above equation dominates and $f(x)$ diverges, signaling that there is no state localized in the well at $E = 0$.

As we now increase E from zero, the symmetric wave function, having zero slope at the centre of the well, will begin to curve downwards on moving away from the centre, and as the energy increases so does the downward curvature.

This naturally changes the mix of increasing and decreasing exponentials needed to connect smoothly at the wall, and in fact as a function of E, $A(E)$ changes sign at a certain value we will call E_0. For energies just below E_0, $f(x)$ diverges to plus infinity for large x. For energies above E_0, it diverges to minus infinity for large x. *Exactly at E_0, $f(x)$ goes to zero for large x.* This is the wave function we are looking for: it corresponds to a particle localized close to the well, and in fact is the lowest possible energy—the *ground state*—for a particle in the well. E_0 is called the ground state *eigenvalue*, the wave function is called an *eigenstate*.

FINDING THE GROUND STATE ENERGY

The first problem is to find this value E_0. If we just guess a value of E, the wave function will almost certainly diverge for large x. The way to find E_0 is to notice that for E below E_0, $f(x)$ goes to large positive values on the far right, for E just above E_0, $f(x)$ goes to large negative values on the far right.

So we take an increasing set of E values starting near zero, and watch for the tail to wag! When this happens, we back up half way, then back or forward as necessary, choosing a set of E values that bracket E_0.

A time or two to get a feeling for how the wave functions behave. A general feature to check is that for $E > V(x)$ they are oscillatory, sine or cosine like, although with changing wavelengths in general; whereas for $E < V(x)$ they have exponential character. You can see why this is by looking at the equation, but it's worth bearing in mind as you examine the curves.

AUTOMATING THE PROCESS

Now to a bit of automation. The strategy we are following is to adjust E until the function $f(x)$ goes to zero for large x, that is, far down at the bottom of our column having several hundred entries. Since we would like to know how well we are doing, it's convenient to have the spreadsheet copy that value to some cell near the top. So, if the end of our column of $f(x_i)$'s is F600, say, then we enter =F600 in some convenient square near the top, say B22. Now we can keep an eye on the distant value as we adjust E.

The curve below was generated by this procedure. It is worth noting that there is a definite probability of finding the particle inside the wall, that is to say, the wave function is nonzero there (and, it turns out, for reasonable, physical values of the parameters):

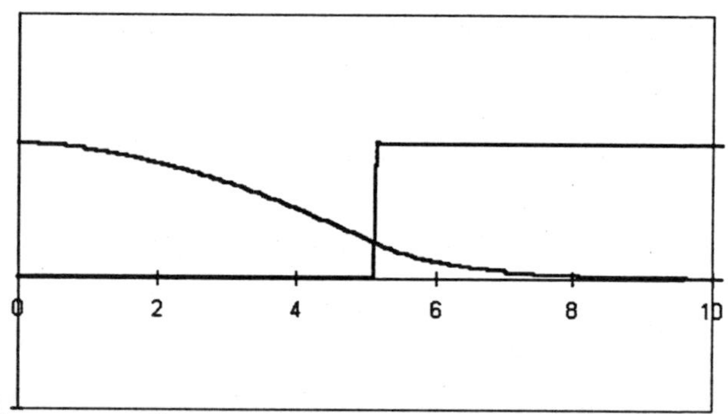

All square well potentials in one dimension, however shallow, have a localized ground state with this general shape. Whether or not there are other eigenstates with other eigenvalues depends on the depth of the potential. For a sufficiently shallow potential, there is only one state.

An infinitely deep well, as we discussed earlier, has an infinite number of bound states. As the well depth increases from zero, states are bound sequentially. These higher eigenstates are called *excited states*. A particle in one of them will usually decay to a lower state, emitting a photon, just as in the Bohr atom.

FINDING EXCITED STATES

Let us now see how the wave function develops as E increases beyond E_0 for a sufficiently deep well. As the energy is increased, the cosine term inside the well has tighter curvature, and the exponentially increasing term for large x is large and negative.

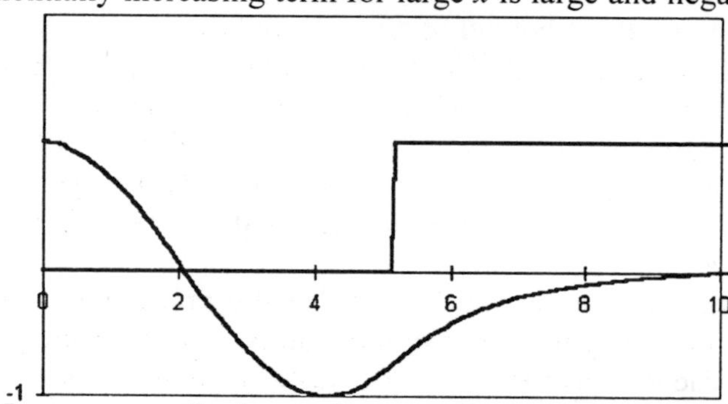

However, continuing to increase the energy, eventually the cosine term bottoms out inside the well and begins to turn up again before reaching the wall. At a certain energy, it again becomes possible to match it to a decreasing function: A point to notice about the wave function pictured above is that has a node at about $x = 2$. (Actually, of course, this is just *half* the wave function for the complete well-there will be another node at $x = -2$.)

As we continue to increase E, the high-x tail of the wave function wags up and down. Each time it crosses the axis there is an allowable wave function. Furthermore, each time it crosses the axis the wave function collects another node for $x > 0$. Thus, the complete wave functions generated by this method have 0, 2, 4, 6, … nodes. From the symmetry of the problem, the allowable wave functions with an *odd* number of nodes must have one node at the origin. They can be generated by taking as initial conditions that f is zero at the origin, and has finite slope.

CURVATURE OF WAVE FUNCTIONS

Schrodinger's equation in the form

$$\frac{d^2 f(x)}{dx^2} = (V(x) - E)f(x)$$

can be interpreted by saying that the left-hand side, the rate of change of slope, is the *curvature*—so the curvature of the function is equal to $(V(x) - E)f(x)$. This means that if $E > V(x)$, for $f(x)$ positive $f(x)$ is curving negatively, for $f(x)$ negative $f(x)$ is curving positively. *In both cases, f(x) is always curving towards the axis.* This means that for $E > V(x)$, $f(x)$ has a kind of stability: its curvature is always bringing it back towards the axis, so it has oscillatory character. On the other hand, for $V(x) > E$, the curvature is always *away* from the axis. This means that $f(x)$ tends to diverge to infinity. Only under exactly the right conditions will this curvature be just enough to bring the wave function to zero as x goes to infinity. (As $f(x)$ tends to zero, the curvature tends to zero, too.)

It is worth examining the wave functions generated by the spreadsheet to see just how the curvature changes as $V(x) - E$, or for that matter $f(x)$, changes sign.

THE SIMPLE HARMONIC OSCILLATOR

A nonrelativistic particle in a potential $\frac{1}{2}Cx^2$, is a system with wide application in both classical and quantum physics. The simplest model is a mass sliding backwards and forwards on a frictionless surface, attached to a fixed wall by a spring, the rest position defined by the natural length of the spring.

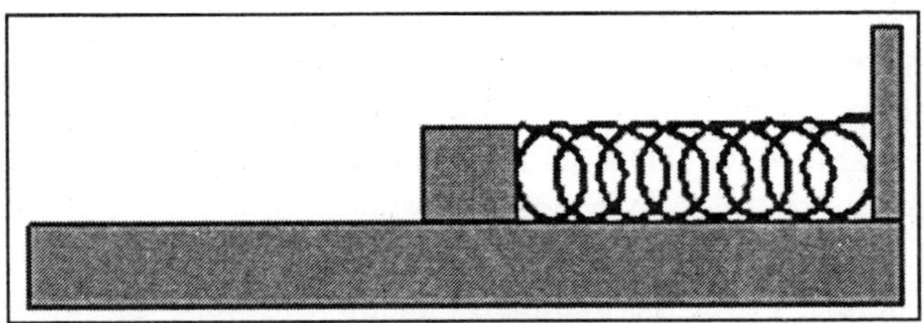

Fig. Spring

Many of the mechanical properties of a crystalline solid can be understood by visualizing it as a regular array of atoms, a cubic array in the simplest instance, with nearest neighbors connected by springs (the valence bonds) so that an atom in a cubic crystal has six such springs attached, parallel to the *x*, *y*, and *z* axes. Provided the oscillations of the atoms are not too large, the springs behave well, and the atom sees itself in a potential

$$\frac{1}{2}kr^2 = \frac{1}{2}kx^2 + \frac{1}{2}ky^2 + \frac{1}{2}kz^2.$$

Now, as the solid is heated up, it should be a reasonable first approximation to take all the atoms to be jiggling about independently, and classical physics, the "Equipartition of Energy", would then assure us that at temperature T each atom would have on average energy $3kT$, k being Boltzmann's constant. The specific heat per atom would then be just $3k$.

But this is *not* what is observed! The specific heats of all solids drop *dramatically* at low temperatures. What's going on here? It took Einstein to figure it out. Black Body Radiation that at low temperatures the blue modes were frozen out because energy could only be absorbed or emitted in quanta, photons, and the energy per quantum was directly proportional to the frequency, so only relatively low energy oscillators gained energy at low temperatures.

Einstein realized that exactly the same considerations must apply to mechanical oscillators, such as atoms in a solid. He assumed each atom to be an independent simple harmonic oscillator, and, just as in the case of black body radiation, the oscillators can only absorb energies in quanta. Consequently, at low enough temperatures there is rarely sufficient energy in the ambient thermal excitations to excite the oscillators, and they freeze out, just like blue oscillators in low temperature black body radiation.

Einstein's picture was later somewhat refined—the basic set of oscillators was taken to be standing sound wave oscillations in the solid rather than individual atoms (even more like black body radiation in a cavity) but the main conclusion was not affected. In the more modern picture of sound waves in a solid, the "elementary" sound wave, analogous to the photon, is called the phonon, and has energy hf, where h is again Planck's constant, and f is the sound frequency.

Oscillations of molecules can usually be analyzed fairly accurately as simple harmonic oscillations, in particular the diatomic molecule. Of course, this picture breaks down for sufficiently large amplitude oscillations—eventually any molecule breaks up.

What kind of wave function do we expect to see in a harmonic oscillator potential? Whatever kinetic energy we give the particle, if it gets far enough from the origin the potential energy will win out, and the wave will decay for the particle going further out. We know that when a particle penetrates a barrier of height V_0, say, greater than the particle's kinetic energy, the wave function decreases exponentially into the barrier, like $e^{-\alpha x}$, where $\alpha = \sqrt{2m(V_0 - E)/\hbar^2}$.

But the simple harmonic oscillator potential is less penetrable than a flat barrier, because its height increases as x^2 as the particle penetrates, so we can see from the expression for α above that for large x α itself increases linearly in x. Of course, this is something of a handwaving argument, the solution of a differential equation for a varying potential is not just a smooth sequence of solutions for constant potentials, but it does suggest that the right wavefunction for the oscillator potential might decay as $e^{-(\text{constant})x^2}$. We write it as $e^{-\frac{x^2}{2\alpha^2}}$, so that the probability distribution is proportional to $e^{-\frac{x^2}{\alpha^2}}$, and *a*, which has the dimensions of length, is a natural measure of the spread of the wave function.

The Schrodinger equation for the simple harmonic oscillator is

$$-\frac{\hbar^2}{2m}\frac{d^2\psi(x)}{dx^2}+\frac{1}{2}Cx^2\psi(x)=E\psi(x).$$

If $\psi(x)=e^{-\frac{x^2}{2\alpha^2}}$, it is straightforward to verify that

$$\frac{d^2\psi}{dx^2}=\frac{1}{a^2}\psi+\frac{x^2}{a^2}\psi$$

Substituting this value inSchrodinger 's equation we find

$$-\frac{\hbar^2}{2m}\left(-\frac{1}{a^2}+\frac{x^2}{a^2}\right)\psi(x)\frac{1}{2}Cx^2\psi(x)=E\psi(x)$$

This equation can only be true for all x if the x^2 terms are separately identically zero, that is,

$$-\frac{\hbar^2}{2ma^4}=\frac{C}{2}, \text{ so } a=\left(\frac{\hbar^2}{mC}\right)^{\frac{1}{4}}.$$

This fixes the wave function. Requiring the remaining terms to balance fixes the energy:

$$E=\frac{\hbar^2}{2ma}=\frac{\hbar^2}{2m}.\frac{\sqrt{mC}}{\hbar}=\frac{1}{2}\hbar\sqrt{\frac{C}{m}}=\frac{1}{2}\hbar\omega_0,$$

where ω_0 is the classical oscillator frequency—given the particle mass m and the spring constant C, the classical equation of motion of the oscillator is

$$m\frac{d^2x}{dt^2}=-Cx.$$

Taking a solution of the form

$$x=x_0\sin\omega_0 t,$$

gives $\omega_0=\sqrt{\frac{C}{m}}$.

An important point here is that the energy is nonzero, just as it was for the square well. The central part of the wave function must have some curvature to join together the decreasing wave function on the left to that on the right. This "zero point energy" is sufficient in one case to melt the lattice helium is liquid even down to absolute zero temperature (checked down to microkelvins!) because of this wave function spread.

TIME DEPENDENT STATES OF THE SIMPLE HARMONIC OSCILLATOR

Working with the time independentSchrodinger equation, as we have in the above, implies that we are restricting ourselves to solutions of the fullSchrodinger equation which have a particularly simple time dependence, an overall phase factor $\varphi(t) = e^{-iBt/k}$, and are states of definite energy E.

However, the full time dependentSchrodinger equation is a linear equation, so if $\psi_1(x,t)$ and $\psi_2(x,t)$ are solutions, so is any linear combination $A\psi_1+B\psi_2$. Assuming ψ_1 and ψ_2 are definite energy solutions for different energies E_1 and E_2, the combination will not correspond to a definite energy a measurement of the energy will give either E_1 or E_2, with appropriate probabilities. In the jargon, the combination is not an "eigenstate" of the energy but it is still a perfectly good, physically realizable wave function.

$$\psi(x, t) = Ae^{-\frac{x^2}{2\alpha^2}} e^{-iB_0 t/\hbar} = Ae^{-\frac{x^2}{2\alpha^2}} e^{-i\omega_0 t/2}$$

It is instructive to examine a combination state of this form a little more closely. We know that for the ground state wave function,
and for the first excited state,

$$\psi(x, t) = Bxe^{-\frac{x^2}{2\alpha^2}} e^{-iB_1 t/\hbar} = Bxe^{-\frac{x^2}{2\alpha^2}} e^{-i3\omega_0 t/2}.$$

Suppose we simply add terms of this type together (neglecting the overall normalization constant for now), for example

$$\psi_{\text{comb}}(x, t) = e^{-\frac{x^2}{2\alpha^2}} e^{-i\omega_0 t/2} = xe^{-\frac{x^2}{2\alpha^2}} e^{-i3\omega_0 t/2}.$$

Looking at this wave function for $t = 0$, we notice that the two terms have the same sign for $x > 0$, and opposite signs for $x < 0$. Therefore, sketching the probability distribution for the particle's position, it is heavily skewed to the right (positive x). However, the two terms have different time-dependent phases, differing by a factor $e^{-i\omega_0 t}$, so after time π/ω_0 has elapsed, a factor of -1 has evolved between the terms.

If we *now* look at the probability distribution $|\psi|^2$, it will be skewed to the left. In other words, if the state is not of definite energy, the probability distribution can vary in time. Of course, the *total* probability of finding the particle *somewhere* stays the same. Note that the probability distribution swings back and forth with the period of the oscillator. This discussion also implies that an ordinary pendulum, which clearly swings back and forth, cannot be in a state of definite energy!

THE THREE DIMENSIONAL SIMPLE HARMONIC OSCILLATOR

It is very simple to go from the one dimensional to the three dimensional simple harmonic oscillator, because the potential $\frac{1}{2}kr^2 = \frac{1}{2}kx^2 + \frac{1}{2}ky^2 + \frac{1}{2}kz^2$ is a sum of separate x, y, z potentials, and consequently any product $\psi(x, y, z) = f(x)\, g(y)\, h(z)$ of three solutions of the one- dimensional harmonic oscillator time independent Schrodinger equation will be a solution of the three-dimensional harmonic oscillator, with energy the sum of the three one-dimensional energies.

So the states are labeled with three quantum numbers, one for each direction, each can be 0, 1, 2, ... If we call these three quantum numbers n_x, n_y, n_z then from what we already know about the one dimensional case, the energy of the three dimensional state must be $\left(n_n + \frac{1}{2}\right)\hbar\omega_0 + \left(n_y + \frac{1}{2}\right)\hbar\omega_0 + \left(n_z + \frac{1}{2}\right)\hbar\omega_0$.

For example, the lowest energy state of the three dimensional harmonic oscillator, the zero point energy, is $\frac{3}{2}\hbar\omega_0$. Obviously, the higher energy states are very degenerate many sets of quantum numbers correspond to the same state because the energy only depends on the *sum* of the three integer quantum numbers. Note that this degeneracy arises from the *symmetry* of the potential, the spring constant k is the same in all three directions. If the potential were of the form $\frac{1}{2}k_x x^2 = \frac{1}{2}k_y y^2 + \frac{1}{2}k_z z^2$ for general k's, there would be no degeneracy. (Such potentials approximately describe oscillations of an atom in an anisotropic crystal.)

Another approach to the three dimensional symmetric $\frac{1}{2}kr^2$ simple harmonic oscillator is to try a separable wave function in spherical polar coordinates, $\psi(r\ \theta, \varphi) = R(r)\ \Theta\ (\theta)\Phi(\varphi)$. This approach is covered in detail in later courses in quantum mechanics, and is the standard method for treating the hydrogen atom (where the potential cannot be written as a sum of x, y, and z potentials).

The angular functions describe the angular momentum of the particle. Some insight can be gained by considering the two dimensional case. Consider a pendulum swinging in the x direction (z is vertical). Now give it a kick so it also has swing in the y direction. In general, it will follow an elliptical path in the x, y plane. The right kick will make it a circle. For the circular orbit, the old fashioned Bohr quantization of angular momentum can be used to find the energy levels.

Chapter 4

Particles in Two-Dimensional

SEPARATION OF VARIABLES IN ONE DIMENSION

We learned from solvingSchrodinger 's equation for a particle in a *one*-dimensional box that there is a set of solutions, the stationary states, for which the time dependence is just an overall rotating phase factor, and these solutions correspond to definite values of the energy. An alternative way of finding that set of solutions is *separation of variables*. The basic strategy is to assume that the solution to the wave equation can be factored into a product of two functions, one depending only on time, the other on the spatial variable,

$$\psi(x, t) = \psi(x)\,\varphi(t).$$

If this solution is substituted in theSchrodinger equation, and the result divided by $\psi(x, t)$, we find

$$\frac{i\hbar\dfrac{\partial\varphi(t)}{\partial t}}{\varphi(t)} = \frac{-\dfrac{\hbar}{2m}\dfrac{\partial\psi(x)}{\partial x}+V(x)\psi(x)}{\psi(x)}$$

On writing the equation in this form, it is clear that the left hand side is only a function of t, not of x, and the right hand side is only a function of x! This can only make sense if in fact both sides are the same constant. If we denote this constant by E, we can write two equations:

$$i\hbar\frac{\partial\varphi(t)}{\partial t} = E\varphi(t)$$

$$-\frac{\hbar^2}{2m}\frac{\partial^2\psi(x)}{\partial x^2}+V(x)\psi(x) = E\psi(x).$$

The solution to the first equation just gives the phase time dependence,

$$\varphi(t) = Ae^{\frac{-iBt}{\hbar}}$$

and the second is the time independentSchrodinger equation as before. The solutions to this equation are determined by the boundary conditions on ψ, in general there is a sequence of such eigenstates labeled by a quantum number $n = 0,1,2,3,\ldots$, with corresponding values E_0, E_1, ..., which are put in the corresponding φ.

A TWO DIMENSIONAL BOX

Let us now consider theSchrodinger equation for an electron confined to a two dimensional box, $0 < x < a$, $0 < y < b$. That is to say, within this rectangle the electron wave function behaves as a free particle ($V(x,y) = 0$), but the walls are impenetrable so the wave function $\psi(x, y, t) = 0$ at the walls. What do we expect the wave function to look like?

First notice that the separation of variables trick given above for one dimension works equally well here, writing $\psi(x, y, t) = \psi(x, y)\,\varphi(t)$ gives

$$i\hbar\frac{\partial\varphi(t)}{\partial t} = E\varphi(t)$$

$$-\frac{\hbar^2}{2m}\left(\frac{\partial^2\psi(x,y)}{\partial x^2} + \frac{\partial^2\psi(x,y)}{\partial y^2}\right) = E\psi(x,y).$$

The surprising thing at this point is that we can do the separation of variables trick *again* we can write $\psi(x, y) = f(x)\,g(y)$ and substitute in the above equation to find

$$-\frac{\hbar^2}{2m}\left(\frac{\frac{\partial^2 f(x)}{\partial x^2}}{f(x)} + \frac{\frac{\partial^2 g(x)}{\partial y^2}}{g(y)}\right) = E.$$

Again, we have an equation in which only one term is x-dependent, so it must be a constant (which we take to be negative for future convenience),

$$\frac{\frac{\partial^2 f(x)}{\partial x^2}}{f(x)} = -C\text{, say, so} \frac{\partial^2 f(x)}{\partial x^2} = -Cf(x).$$

This is exactly the same equation we dealt with in the one dimensional case, so we know

$$f(x) = A \sin \frac{n\pi x}{a}$$

with n an integer, and the constant C is equal to $n^2\pi^2/a^2$. Hence the energy levels in this rectangular well are given by

$$E_{n,m} = \frac{\hbar^2}{2m}\left(\frac{n^2\pi^2}{a^2} + \frac{m^2\pi^2}{b^2}\right)$$

with n, m the two quantum numbers needed to label each state.

DEGENERATE STATES

Two distinct wave functions are said to be *degenerate* if they correspond to the same energy. If the sides a, b of the rectangle are such that a/b is irrational (the general case), there will be no degeneracies. The *most* degenerate case is the square, $a = b$, for which clearly $E_{m,n} = E_{n,m}$. Degeneracies in quantum physics are most often associated with symmetries in this way.

We give here examples of wave functions (3,2) and (2,3) for a rectangle. These are contour maps for the time-independent solution, with white being the highest point. These two wave functions do not correspond to the same energy, although they would, of course, for a square.

A TWO DIMENSIONAL CIRCULAR WELL

There is a well-known (scanning tunneling microscope) picture of a "corral" of 48 iron atoms arranged in a circle on a flat surface. The picture suggests that there are local electron states in the corral, there are concentric rings of electron density inside (but not outside) the atomic ring.

Let us assume the situation is well described by $V(r) = \infty$ for $r < a$, $V(r) =$ for $r > a$. It is convenient to write the time independentSchrodinger equation in the form

$$(\nabla^2 + k^2)\,\psi = 0, \text{ where } k^2 = \frac{2mE}{\hbar^2}.$$

Since the system we are analyzing has circular symmetry, it is natural to try for a solution by separating variables

$$\psi(r, \theta) = R(r)\,\Theta\,(\theta).$$

Expressing ∇^2 in (r,θ) coordinates,Schrodinger 's equation in zero potential becomes:

$$\frac{1}{r}\frac{\partial}{\partial r}\left(r\frac{\partial\psi}{\partial r}\right)+\frac{1}{r^2}\frac{\partial^2\psi}{\partial\theta^2}+k^2\psi = 0.$$

Substituting the factored wave function $\psi(r, \theta) = R(r)\,\Theta\,(\theta)$ in this equation, and dividing by ψ as before, gives the two equations:

$$\frac{\partial^2\Theta(\theta)}{\partial\theta^2} = m^2\Theta(\theta)$$

$$\frac{\partial^2 R(r)}{\partial r^2}+\frac{1}{r}\frac{dR(r)}{\partial r}+\left(k^2-\frac{m^2}{r^2}\right)R(r) = 0.$$

The constant term that arises when a function of r is equal to a function of θ we set equal to m^2, for reasons that are immediately apparent on examining the θ-equation above. The solutions to this equation are just $e^{im\theta}$ and $e^{-im\theta}$, will be single valued only if m is an integer.

Thus the quantum number m gives the number of oscillations of the wave function on going in a path around the origin. This is strongly reminiscent of the de Broglie interpretation of the Bohr quantum condition.

The equation for the radial function $R(r)$ is actually a well known equation in mathematical physics, it is Bessel's equation. (The two equations together describe vibrations of a circular drumhead, or waves on the surface of a cup of coffee.) We must then solve Bessel's equation subject to the requirements that the function vanish at $r = a$ (the wall) and also that it be finite at $r = 0$ (the $1/r^2$ term in the last bracket means that not all solutions of Bessel's equation satisfy this condition). Fortunately, Bessel's equation has been thoroughly explored, and we can just look up solutions.

CLASSICAL WAVE EQUATIONS

A fairly brief review of waves in various shaped elastic media—beginning with a taut string, then going on to an elastic sheet, a drumhead, first of rectangular shape then circular, and finally considering elastic waves on a spherical surface, like a balloon.

The reason we look at this material here is that these are "real waves", hopefully not too difficult to think about, and yet mathematically they are the solutions of the same wave equation theSchrodinger wave function obeys in various contexts, so should be helpful in visualizing solutions to that equation, in particular for the hydrogen atom.

We begin with the stretched string, then go on to the rectangular and circular drumheads. We derive the wave equation from $F = ma$ for a little bit of string or sheet. The equation corresponds exactly to theSchrodinger equation for a free particle with the given boundary conditions.

The most important section here is the one on waves on a sphere. We find the first few standing wave solutions. These waves correspond toSchrodinger 's wave function for a free particle on the surface of a sphere. This is what we need to analyse to understand the hydrogen atom, because using separation of variables we split the electron's motion into radial motion and motion on the surface of a sphere.

The potential *only* affects the *radial* motion, so the motion on the sphere is free particle motion, described by the same waves we find for vibrations of a balloon. (There *is* the generalization to complex non-standing waves, parallel to the one-dimensional extension from sinkx and coskx to e^{ikx} and e^{-ikx}, but this does not affect the structure of the equations.)

WAVES ON A STRING

Let's begin by reminding ourselves of the wave equation for waves on a taut string, stretched between $x = 0$ and $x = L$, tension T newtons, density ρ kg/meter. Assuming the string's equilibrium position is a straight horizontal line (and, therefore, ignoring gravity), and assuming it oscillates in a vertical plane, we use $f(x,t)$ to denote its shape at instant t, so $f(x,t)$ is the instantaneous upward displacement of the string at position x. We assume the amplitude of oscillation remains small enough that the string tension can be taken constant throughout.

The *wave equation* is derived by applying $F = ma$ to an infinitesimal length dx of string. We picture our little length of string as bobbing up and down in simple harmonic motion, which we can verify by finding the net force on it as follows.

At the left hand end of the string fragment, point x, say, the tension T is at a small angle $df(x)/dx$ to the horizontal, since the tension acts necessarily along the line of the string. Since it is pulling to the left, there is a *downward* force component $Tdf(x)/dx$. At the right hand end of the string fragment there is an *upward* force $Tdf(x + dx)/dx$.

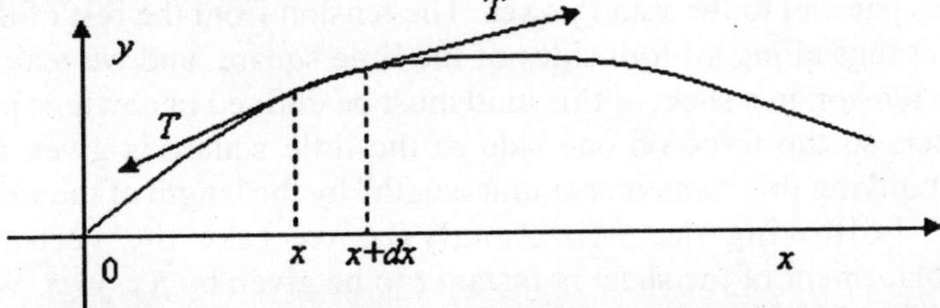

Putting $f(x + dx) = f(x) + (df/dx)dx$, and adding the almost canceling upwards and downwards forces together, we find a net force $T(d^2f/dx^2)dx$ on the bit of string. The string mass is $\rho\, dx$, so $F = ma$ becomes

$$T\frac{\partial^2 f(x,t)}{\partial x^2}dx = \rho dx\frac{\partial^2 f(x,t)}{\partial t^2}$$

giving the standard wave equation

$$\frac{\partial^2 f(x,t)}{\partial x^2} = \frac{1}{c^2}\frac{\partial^2 f(x,t)}{\partial t^2}$$

with wave velocity given by $c^2 = T/\rho$. (A more detailed discussion is given in my Physics 152 Course, plus an animation here.)

This equation can of course be solved by separation of variables, $f(x,t) = f(x)g(t)$, and the equation for $f(x)$ is identical to the time independentSchrodinger equation for a particle confined to $(0, L)$ by infinitely high walls at the two ends. This is why the eigenfunctions (states of definite energy) for aSchrodinger particle confined to $(0, L)$ are identical to the modes of vibration of a string held between those points. (However, it should be realized that the *time dependence* of the string wave equation and theSchrodinger *time-dependent* equation are quite different, so a nonstationary state, one corresponding to a sum of waves of different energies, will develop differently in the two systems.)

WAVES ON A RECTANGULAR DRUMHEAD

Let us now move up to two dimensions, and consider the analogue to the taut string problem, which is waves in a taut horizontal elastic sheet, like, say, a drumhead. Let us assume a rectangular drumhead to begin with. Then, parallel to the argument above, we would apply $F = ma$ to a small square of elastic with sides parallel to the x and y axes. The tension from the rest of the sheet tugs along all four sides of the little square, and we realise that tension in a sheet of this kind must be defined in newtons per meter, so the force on one side of the little square is given by multiplying this "tension per unit length" by the length of the side.

Following the string analysis, we take the vertical displacement of the sheet at instant t to be given by $f(x, y, t)$. We assume this displacement is quite small, so the tension itself doesn't vary, and that each bit of the sheet oscillates up and down (the sheet is not tugged to one side).

Suppose the bottom left-hand corner (so to speak) of the square is (x, y), the top right-hand corner $(x + dx, y + dy)$. Then the left and right edges of the square have lengths dy. Now, what is the total force on the left edge? The force is Tdy, in the local plane of the sheet, perpendicular to the edge dy. Factoring in the slope of the sheet in the direction of the force, the vertically downward component of the force must be $Td\partial f(x,y,t)/\partial x$. By the same argument, the force on the right hand edge has to have an *upward* component $Tdy\, \partial f(x+dx, y, t)/\partial x$.

Thus the net upward force on the little square from the sheet tension tugging on its left and right sides is

$$Tdy\left(\frac{\partial f(x, dx, y, t)}{\partial x} - \frac{\partial f(x, dx, y, t)}{\partial x}\right) = Tdydx\left(\frac{\partial^2 f}{\partial x^2}\right)$$

The net vertical force from the sheet tension on the other two sides is the same with *x* and *y* interchanged. The mass of the little square of elastic sheet is $\rho dxdy$, and its upward acceleration is $\partial^2 f/\partial t^2$. Thus $F = ma$ becomes:

$$Tdydx\left(\frac{\partial^2 f}{\partial x^2}\right) + Tdydx\left(\frac{\partial^2 f}{\partial y^2}\right) = \rho dxdy\left(\frac{\partial^2 f}{\partial t^2}\right)$$

giving

$$\left(\frac{\partial^2 f}{\partial x^2}\right) + \left(\frac{\partial^2 f}{\partial y^2}\right) = \frac{1}{c^2}\left(\frac{\partial^2 f}{\partial t^2}\right)$$

with $c^2 = T/\rho$. This equation can be solved by separation of variables, and the time independent part is identical to theSchrodinger time independent equation for a free particle confined to a rectangular box.

WAVES ON A CIRCULAR DRUMHEAD

A similar argument gives the wave equation for a circular drumhead, this time in (r, φ) coordinates (we use φ rather than θ here because of its parallel role in the spherical case, to be discussed shortly).This time, instead of a tiny square of elastic, we take the small area $rdrd\varphi$ bounded by the circles of radius r and $r + dr$ and lines through the origin at angles φ and $\varphi + d\varphi$. Now, the downward force from the tension T in the sheet on the inward curved edge, which has length $rd\varphi$, is $Trd\varphi\ f(r, \varphi, t)/\partial r$.

On putting this together with the upward force from the other curved edge, it is important to realise that the r in $Trd\varphi$ varies as well as $\partial f/\partial r$ on going from r to $r + dr$, so the sum of the two terms is $Td\varphi\partial/\partial r(r\partial f/\partial r)dr$. To find the vertical elastic forces from the straight sides, we need to find how the sheet slopes in the direction perpendicular to those sides. The measure of length in that direction is not φ, but $r\varphi$, so the slope is $1/r.\partial f/\partial\varphi$, and the net upward elastic force contribution from those sides (which have length dr) is $Tdrd\varphi\partial/\partial\varphi\ (1/r.\partial f/\partial\varphi)$.

Writing $F = ma$ for this small area of elastic sheet, of mass $\rho rdrd\varphi$, gives then

$$Td\varphi\frac{\partial}{\partial r}r\frac{\partial f}{\partial r}dr + Tdrd\varphi\frac{\partial}{\partial\varphi}\frac{1}{r}\frac{\partial f}{\partial\varphi} = \rho rdrd\varphi\frac{\partial f}{\partial t^2}.$$

which can be written

$$\frac{1}{r}\frac{\partial}{\partial r}r\frac{\partial f}{\partial r}+\frac{1}{r^2}\frac{\partial^2 f}{\partial \varphi^2}=\frac{1}{c^2}\frac{\partial^2 f}{\partial \varphi t}.$$

This is the wave equation in polar coordinates. Separation of variables gives a radial equation called Bessel's equation, the solutions are called Bessel functions. The corresponding electron standing waves have actually been observed for an electron captured in a circular corral on a surface.

WAVES ON A SPHERICAL BALLOON

Finally, let us consider elastic waves on the surface of a sphere, such as an inflated spherical balloon. The natural coordinate system here is spherical polar coordinates, with θ measuring latitude, but counting the north pole as zero, the south pole as π. The angle φ measures longitude from some agreed origin.

We take a small elastic element bounded by longitude lines φ and $\varphi + d\varphi$ and latitude θ and $\theta + d\theta$. For a sphere of radius *r*, the sides of the element have lengths $r\sin\theta\, d\varphi$, $rd\theta$ etc. Beginning with one of the longitude sides, length $rd\theta$, tension *T*, the only slightly tricky point is figuring its deviation from the local horizontal, which is $1/r\sin\theta.(\partial f/\partial\varphi)$, since increasing φ by $d\varphi$ means moving an actual distance $r\sin\theta\, d\varphi$ on the surface, just analogous with the circular case above. Hence, by the usual method, the actual vertical force from tension on the two longitude sides is $Trd\theta\, d\varphi. (\partial/\partial\varphi)1/r\sin\theta.(\partial f/\partial\varphi)$.

To find the force on the latitude sides, taking the top one first, the slope is given by $1/r.\partial f/\partial\theta$, so the force is just $Tr\sin\theta\, d\varphi.1/r.\partial f/\partial\theta$. On putting this together with the opposite side, it is necessary to recall that $\sin\theta$ as well as *f* varies with θ, so the sum is given by: $Trd\varphi d\theta\partial/\partial\theta \sin\theta.1/r.\partial f/\partial\theta$. We are now ready to write down $F = ma$ once more, the mass of the element is $\rho r^2\sin\theta\, d\theta\, d\varphi$. Canceling out elements common to both sides of the equation, we find:

$$\frac{1}{\sin\theta}\frac{\partial}{\partial\theta}\sin\theta\frac{\partial f}{\partial\theta}+\frac{1}{\sin^2\theta}\frac{\partial^2 f}{\partial\varphi^2}=\frac{1}{c^2}\frac{\partial^2 f}{\partial t^2}.$$

Again, this wave equation is solved by separation of variables.

The time-independent solutions are called the *Legendre* functions. They are the basis for analyzing the vibrations of any object with spherical symmetry, for example a planet struck by an asteroid, or vibrations in the sun generated by large solar flares.

SIMPLE SOLUTIONS TO THE SPHERICAL WAVE EQUATION

Recall that for the two dimensional circular case, after separation of variables the angular dependence was all in the solution to $\partial^2 f/\partial\varphi^2 = -\lambda f$, and the physical solutions must fit smoothly around the circle (no kinks, or it would not satisfy the wave equation at the kink), leading to solutions $\sin m\varphi$ and $\cos m\varphi$ (or $e^{im\varphi}$) with m an integer, and $\lambda = m^2$ (this is why we took λ with a minus sign in the first equation).

For the spherical case, the equation containing all the angular dependence is

$$\frac{1}{\sin\theta}\frac{\partial}{\partial\theta}\sin\theta\frac{\partial f}{\partial\theta}+\frac{1}{\sin^2\theta}\frac{\partial^2 f}{\partial\varphi^2}=-\lambda f$$

The standard approach here is, again, separation of variables. Taking the first term on the left hand side over to the right, and multiplying throughout by $\sin^2\theta$ isolates the φ term:

$$\frac{\partial^2 f}{\partial\varphi^2}=-\sin^2\theta\left(\lambda f+\frac{1}{\sin\theta}\frac{\partial}{\partial\theta}\sin\theta\frac{\partial f}{\partial\theta}\right)$$

Writing now

$$f(\theta,\varphi)=f_\theta(\theta)f_\varphi(\varphi)$$

in the above equation, and dividing throughout by f, we find as usual that the left hand side depends only on φ, the right hand side only on θ, so both sides must be constants. Taking the constant as $-m^2$, the φ solution is $e^{\pm im\varphi}$, and one can insert that in the θ equation to give

$$\frac{1}{\sin\theta}\frac{\partial}{\partial\theta}\sin\theta\frac{\partial f_\theta}{\partial\theta}+\frac{m^2}{\sin^2\theta}f_\theta=-\lambda f_\theta$$

What about possible solutions that don't depend on φ? The equation would be the simpler

$$\frac{1}{\sin\theta}\frac{\partial}{\partial\theta}\sin\theta\frac{\partial f}{\partial\theta} = -\lambda f$$

Obviously, f = constant is a solution (for $m = 0$) with eigenvalue $\lambda = 0$.

Try $f = \cos\theta$. It is easy to check that this is a solution with $\lambda = 2$.

Try $f = \sin\theta$. This is *not* a solution. In fact, we should have realized it cannot be a solution to the wave equation by visualizing the shape of the elastic sheet near the north pole. If $f = \sin\theta$, $f = 0$ at the pole, but rises linearly (for small θ) going away from the pole. Thus the pole is at the bottom of a conical valley. But this conical valley amounts to a kink in the elastic sheet—the slope of the sheet has a discontinuity if one moves along a line passing through the pole, so the shape of the sheet cannot satisfy the wave equation at that point.

This is somewhat obscured by working in spherical coordinated centered there, but locally the north pole is no different from any other point on the sphere, we could just switch to local (x,y) coordinates, and the cone configuration would clearly *not* satisfy the wave equation.

However, $f = \sin\theta \sin\varphi$ *is* a solution to the equation. It is a worthwhile exercise to see how the φ term gets rid of the conical point at the north pole by considering the value of f as the north pole is approached for various values of φ: $\varphi = 0, \pi/2, \pi, 3\pi/2$ say. The sheet is now smooth at the pole!

We find $f = \sin\theta \cos\varphi$, $\sin\theta \sin\varphi$ (and so $\sin\theta\, e^{i\varphi}$) are solutions with $\lambda = 2$.

It is straightforward to verify that $f = \cos^2\theta - 1/3$ is a solution with $\lambda = 6$.

Finally, we mention that other $\lambda = 6$ solutions are $\sin\theta \cos\theta \sin\varphi$ and $\sin^2\theta \sin 2\varphi$.

We do not attempt to find the general case here, but we have done enough to see the beginnings of the pattern. We have found the series of eigenvalues 0, 2, 6, ... It turns out that the complete series is given by $\lambda = l(l + 1)$, with $l = 0, 1, 2, \ldots$. This integer l is the analogue of the integer m in the wave on a circle case. Recall that for the wave on the *circle*, if we chose real wave functions

($\cos m\varphi$, $\sin m\varphi$, not $e^{im\varphi}$) then $2m$ gave the number of nodes the wave had (that is, m complete wavelengths fitted around the circle). It turns out that on the sphere l gives the number of nodal *lines* (or circles) on the surface.

This assumes that we again choose the φ-component of the wave function to be real, so that there will be m nodal circles passing through the two poles corresponding to the zeros of the $\cos m\varphi$ term. We find that there are $l - m$ nodal *latitude* circles corresponding to zeros of the function of θ.

SUMMARY: FIRST FEW STANDING WAVES ON THE BALLOON

λ	l	m	*Form of solution (unnormalized)*
0	0	0	constant
2	1	0	$\cos\theta$
2	1	1	$\sin\theta\ e^{i\varphi}$
2	1	−1	$\sin\theta\ e^{-i\varphi}$
6	2	0	$\cos^2\theta - 1/3$
6	2	±1	$\cos\theta \sin\theta\ e^{\pm i\varphi}$
6	2	±2	$\sin^2\theta\ e^{\pm 2i\varphi}$

THESCHRODINGER EQUATION FOR THE HYDROGEN ATOM: HOW DO WE SEPARATE THE VARIABLES?

In three dimensions, theSchrodinger equation for an electron in a potential can be written:

$$-\frac{\hbar^2}{2m}\left(\frac{\partial^2\psi}{\partial x^2}+\frac{\partial^2\psi}{\partial y^2}+\frac{\partial^2\psi}{\partial z^2}\right)+V(x,y,z)\psi = E\psi$$

This is the obvious generalization of our previous two-dimensional discussion, and we will later be using the equation in the above form to discuss electron wave functions in metals, where the standard approach is to work with standing waves in a rectangular box.

Recall that in our original "derivation" of theSchrodinger equation, by analogy with the Maxwell wave equation for light waves, we argued that the differential wave operators arose from the energy-momentum relationship for the particle, that is,

$$\frac{p_x^2+p_y^2+p_z^2}{2m}\psi \equiv -\frac{\hbar^2}{2m}\left(\frac{\partial^2\psi}{\partial x^2}+\frac{\partial^2\psi}{\partial y^2}+\frac{\partial^2\psi}{\partial z^2}\right)=\frac{\hbar^2\nabla^2\psi}{2m}$$

so that the time-independentSchrodinger wave equation is nothing but the statement that E = $K.E.$ + $P.E.$ with the kinetic energy expressed as the equivalent operator.

To make further progress in solving the equation, the only trick we know is separation of variables. Unfortunately, this won't work with the equation as given above in (x, y, z) coordinates, because the potential energy term is a function of x, y and z in a nonseparable form.

The solution is, however, fairly obvious: the potential is a function of radial distance from the origin, independent of direction.

Therefore, we need to take as our coordinates the radial distance r and two parameters fixing direction, θ and φ. We should then be able to separate the variables, because the potential only affects radial motion. No potential term will appear in the equations for θ,φ motion, that will be free particle motion on the surface of a sphere.

MOMENTUM AND ANGULAR MOMENTUM WITH SPHERICAL COORDINATES

It is worth thinking about what are the natural momentum components for describing motion in spherical polar coordinates (r, θ, φ). The radial component of momentum, p_r, points along the radius, of course. The θ-component p_θ points along a line of longitude, away from the north pole if positive (remember θ itself measures *latitude*, counting the north pole as zero). The φ-momentum component, p_φ, points along a line of latitude.

It will be important in understanding the hydrogen atom to connect these momentum components $(p_r, p_\theta, p_\varphi)$ with the *angular* momentum components of the atom. Evidently, momentum in the r-direction, which passes directly through the centre of the atom, contributes nothing to the angular momentum.

Consider now a particle for which $p_r = p_\theta = 0$, only p_φ being nonzero. Classically, such a particle is circling the north pole at constant latitude θ, say, so it is moving in space in a circle or radius $r\sin\theta$ in a plane perpendicular to the north-south axis of the sphere. Therefore, it has an angular momentum about that axis

$$p_\varphi r \sin\theta = L_z, \text{ say.}$$

(The standard transformation from (x, y, z) coordinates to (r, θ, φ) coordinates is to take the north pole of the θ, φ sphere to be on the z-axis.)

The wave equation describing the φ motion is a simple one, with solutions of the form $e^{im\varphi}$ with integer m, just as in the two-dimensional circular well. This just means that *the component of angular momentum along the z-axis is quantized,* $L_z = m\hbar$*, with m an integer*.

TOTAL ANGULAR MOMENTUM AND WAVES ON A BALLOON

The *total* angular momentum is $L = rp\perp$, where $p\perp$ is the component of the particle's momentum perpendicular to the radius, so

$$p_\perp^2 = p_\varphi^2 + p_\theta^2 \,.$$

Thus the square of the total angular momentum is (apart from a constant factor) the kinetic energy of a particle moving freely on the surface of a sphere. The equivalentSchrodinger equation for such a particle is the wave equation given in the last section for waves on a balloon. (This can be established by the standard change of variables routine on the differential operators). Therefore, the solutions we found for elastic waves on a sphere actually describe the angular momentum wave function of the hydrogen atom. We conclude that the total angular momentum is quantized, $L^2 = l(l+1)\hbar^2$, with *l an integer.*

ANGULAR MOMENTUM AND THE UNCERTAINLY PRINCIPLE

The conclusions of our above waves on a sphere analysis of the angular momentum of a quantum mechanical particle are a little strange. We found that the component of angular momentum in the z-direction must be a whole number of ' units, yet the square of the total angular momentum $L^2 = l(l+1)\hbar^2$ is not a perfect square! One might wonder if the component of angular momentum in the x-direction isn't also a whole number of ' units as well, and if not, why not? The key is that in questions of this type we are forgetting the essentially wavelike nature of the particle's motion, or, equivalently, the uncertainty principle. Recall first that the z-component of angular momentum, that is, the angular momentum about the z-axis, is the product of the particle's momentum in the xy-plane and the distance of the line of that motion from the origin. There is no contradiction in specifying that momentum and that position simultaneously, because they are in perpendicular directions.

However, we cannot at the same time specify either of the other components of the angular momentum, because that would involve measuring some component of momentum in a direction in which we have just specified a position measurement. We *can* measure the total angular momentum, that involves additionally only the component p_θ of momentum perpendicular to the p_φ needed for the z-component.

Thus the uncertainty principle limits us to measuring at the same time only the total angular momentum and the component in one direction. Note also that if we knew the z-component of angular momentum to be $m\hbar$, and the total angular momentum were $L^2 = l^2\hbar^2$ with $l = m$, then we would also know that the x and y components of the angular momentum were exactly zero. Thus we would know all three components, in contradiction to our uncertainly principle arguments. This is the essential reason why the square of the total angular momentum is greater than the maximum square of any one component. It is as if there were a "zero point motion" fuzzing out the direction.

Another point related to the uncertainty principle concerns measuring just where in its circular (say) orbit the electron is at any given moment. How well can that be pinned down? There is an obvious resemblance here to measuring the position and momentum of a particle at the same time, where we know the fuzziness of the two measurements is related by $\Delta p \Delta x \sim h$. Naovely, for a circular orbit of radius r in the xy-plane, $pr = L_z$ and distance measured around the circle is $r\theta$, so $\Delta p \Delta x \sim \hbar$ suggests $\Delta L_z \Delta\theta \sim h$.

That is to say, precise knowledge of L_z implies no knowledge of where on the circle the particle is. This is not surprising, because we have found that for $L_z = m\hbar$ the wave has the form $e^{im\varphi}$, and so $|\psi|^2$, the relative probability of finding the particle, is the same anywhere in the circle. On the other hand, if we have a time-dependent wave function describing a particle orbiting the nucleus, so that the probability of finding the particle at a particular place varies with time, the particle cannot be in a definite angular momentum state. This is just the same as saying that a particle described by a wave packet cannot have a definite momentum.

THESCHRODINGER EQUATION IN (r, θ, φ) COORDINATES

It is worth writing first the energy equation for a classical particle in the Coulomb potential:

$$\frac{1}{2m}(p_r^2 + p_\theta^2 + p_\varphi^2) - \frac{1}{4\pi\varepsilon_0}\frac{e^2}{r} = E$$

This makes it possible to see, term by term, what the various parts of theSchrodinger equation signify. In spherical polar coordinates,Schrodinger 's equation is:

$$-\frac{\hbar^2}{2m}\left(\frac{1}{r}\frac{\partial^2}{\partial r^2}(r\psi)\frac{1}{r^2}\left\{\frac{1}{\sin\theta}\frac{\partial}{\partial\theta}\left(\sin\theta\frac{\partial\psi}{\partial\theta}\right)+\frac{1}{\sin^2\theta}\frac{\partial^2\psi}{\partial\varphi^2}\right\}\right)$$

$$-\frac{1}{4\pi\varepsilon_0}\frac{e^2}{r}\psi = E\psi$$

SEPARATING THE VARIABLES: THE MESSY DETAILS

We look for separable solutions of the form

$$\psi(r, \theta, \varphi) = R(r)\,\Theta(\theta)\,\Phi(\varphi)$$

We now follow the standard technique. hat is to say, we substitute $R\Theta\Phi$ for ψ in each term in the above equation. We then observe that the differential operators only actually operate on one of the factors in any given part of the expression, so we put the other two factors to the left of these operators. We then divide the entire equation by $R\Theta\Phi$, to get

$$-\frac{\hbar^2}{2m}\left(\frac{1}{R}\frac{1}{r}\frac{\partial^2}{\partial r^2}(rR)\frac{1}{r^2}\left\{\frac{1}{\Theta}\frac{1}{\sin\theta}\frac{\partial}{\partial\theta}\left(\sin\theta\frac{\partial\Theta}{\partial\theta}\right)+\frac{1}{\sin^2\theta}\frac{\partial^2\Phi}{\partial\varphi^2}\right\}\right)$$

$$-\frac{1}{4\pi\varepsilon_0}\frac{e^2}{r}\psi = E$$

SEPARATING OUT AND SOLVING THE $\Phi(\varphi)$ EQUATION

The above equation can be rearranged to give:

$$\left(\frac{1}{R}\frac{1}{r}\frac{\partial^2}{\partial r^2}(rR)\frac{1}{r^2}\left\{\frac{1}{\Theta}\frac{1}{\sin\theta}\frac{\partial}{\partial\theta}\left(\sin\theta\frac{\partial\Theta}{\partial\theta}\right)+\frac{1}{\sin^2\theta}\frac{1}{\Phi}\frac{\partial^2\Phi}{\partial\varphi^2}\right\}\right)$$

$$=\frac{2m}{\hbar^2}\left(E+\frac{1}{4\pi\varepsilon_0}\frac{e^2}{r}\right)$$

Further rearrangement leads to:

$$\frac{1}{\Phi}\frac{\partial^2\Phi}{\partial\varphi^2}=\sin^2\theta\left[\left(r^2\left\{\frac{2m}{\hbar^2}\left(E+\frac{1}{4\pi\varepsilon_0}\frac{e^2}{r}\right)+\frac{1}{R}\frac{1}{r}\frac{\partial^2}{\partial r^2}(rR)\right\}\right)\right.$$

$$\left.-\frac{1}{\Theta}\frac{1}{\sin\theta}\frac{\partial}{\partial\theta}\left(\sin\theta\frac{\partial\Theta}{\partial\theta}\right)\right]$$

At this point, we have achieved the separation of variables! The left hand side of this equation is a function *only of* φ, the right hand side is a function *only of* r *and* θ. The only way this can make sense is if *both sides of the equation are in fact constant* (and of course equal to each other).

Taking the left hand side to be equal to a constant we denote for later convenience by m^2,

$$\frac{\partial^2\Phi(\varphi)}{\partial\varphi^2}=-m^2\Phi(\varphi)$$

We write the constant $-m^2$ because we know that as a factor in a wave function $\Phi(\varphi)$ must be single valued as φ increases through 2π, so an integer number of oscillations must fit around the circle, meaning Φ is sin$m\varphi$, cos$m\varphi$ or $e^{im\varphi}$ with m an integer. These are the solutions of the above equation. Of course, this is very similar to the particle in the circle in two dimensions, m signifies units of angular momentum about the z-axis.

SEPARATING OUT THE Θ(*θ*) EQUATION

Backing up now to the equation in the form

$$\left(\frac{1}{R}\frac{1}{r}\frac{\partial^2}{\partial r^2}(rR)\frac{1}{r^2}\left\{\frac{1}{\Theta}\frac{1}{\sin\theta}\frac{\partial}{\partial\theta}\left(\sin\theta\frac{\partial\Theta}{\partial\theta}\right)+\frac{1}{\sin^2\theta}\frac{1}{\Phi}\frac{\partial^2\Phi}{\partial\varphi^2}\right\}\right)$$

$$=\frac{2m}{\hbar^2}\left(E+\frac{1}{4\pi\varepsilon_0}\frac{e^2}{r}\right)$$

we can replace the $\frac{1}{\Phi}\frac{\partial^2\Phi}{\partial\Phi^2}$ term by $-m^2$, and move the r term over to the right, to give

$$\frac{1}{\Theta}\frac{1}{\sin\theta}\frac{\partial}{\partial\theta}\left(\sin\theta\frac{\partial\Theta}{\partial\theta}\right)-\frac{m^2}{\sin^2\theta}=$$

$$r^2\left\{\frac{2m}{\hbar^2}\left(E+\frac{1}{4\pi\varepsilon_0}\frac{e^2}{r}\right)-\frac{1}{R}\frac{1}{r}\frac{\partial^2}{\partial r^2}(rR)\right\}$$

We have again managed to *separate the variables* the left hand side is a function *only* of θ, the right hand side a function of r. Therefore both must be equal to the same constant, which we set equal to $-\lambda$.

This gives the $\Theta(\theta)$ equation:

$$\frac{1}{\sin\theta}\frac{\partial}{\partial\theta}\sin\frac{\partial\Theta(\theta)}{\partial\theta}-\frac{m^2}{\sin^2\theta}\Theta(\theta)=-\lambda\Theta(\theta)$$

This is exactly the wave equation we discussed above for the elastic sphere, and the allowed eigenvalues λ are $l(l+1)$, where l = 0, 1, 2,.. with $l \geq m$.

THE *R*(*R*) EQUATION

Replacing the θ, φ operator with the value found just above in the originalSchrodinger equation gives the equation for the radial wave function:

$$-\frac{\hbar^2}{2m}\left(\frac{1}{r}\frac{\partial^2}{\partial r^2}(rR(r))-\frac{l(l+1)}{r^2}R(r)\right)-\frac{1}{4\pi\varepsilon_0}\frac{e^2}{r}R(r) = ER(r)$$

The first term in this radial equation is the usual radial kinetic energy term, equivalent to $p_r^2/2m$ in the classical picture. The third term is the Coulomb potential energy. The second term is an effective potential representing the centrifugal force. This is clarified by reconsidering the energy equation for the classical case,

$$\frac{1}{2m}\left(p_r^2+p_\theta^2+p_\varphi^2\right)-\frac{1}{4\pi\varepsilon_0}\frac{e^2}{r} = E$$

The angular momentum squared

$$L^2 = r^2\left(p_\theta^2+p_\varphi^2\right)=l(l+1)\hbar^2.$$

Thus for fixed angular momentum, we can write the above "classical" equation as

$$\frac{1}{2m}\left(p_r^2+\frac{l(l+1)\hbar^2}{r^2}\right)-\frac{1}{4\pi\varepsilon_0}\frac{e^2}{r} = E.$$

The parallel to the radial Schrodinger equation is then clear.

We must find the solutions of the radial Schrodinger equation that decay for large r. These will be the bound states of the hydrogen atom. In natural units, measuring lengths in units of the first Bohr radius, and energies in Rydberg units

$$r = a_0\rho = \frac{4\pi\varepsilon_0\hbar^2}{me^2}\rho, \quad E = E_R\varepsilon\frac{me^4}{2\hbar^2(4\pi\varepsilon_0)^2}\varepsilon.$$

Finally, taking $u(r) = rR(r)$, the radial equation becomes.

$$-\frac{d^2u(\rho)}{d\rho^2}+\frac{l(l+1)}{\rho^2}u(\rho)-\frac{2}{\rho}u(\rho)=\varepsilon u(\rho).$$

Chapter 5

Wavefunctions and Symmetry

TWO ELECTRONS IN A ONE DIMENSIONAL WELL

So far, we have usedSchrodinger 's equation to see how a single particle, usually an electron, behaves in a variety of potentials. If we are going to think about atoms other than hydrogen, it is necessary to extend theSchrodinger equation so that it describes more than one particle.

As a simple example of a two particle system, let us consider two electrons confined to the same one-dimensional infinite square well,

$$V(x) = 0, -L/2 < x < L/2$$
$$V(x) = \infty \text{ otherwise.}$$

To make things even simpler, let us assume that the electrons do not interact with each other—we switch off their electrostatic repulsion. Then by analogy with our construction of theSchrodinger equation for a single electron, we write

$$-\frac{\hbar^2}{2m}\frac{\partial^2\psi(x_1,x_2,t)}{\partial x_1^2}-\frac{\hbar^2}{2m}\frac{\partial^2\psi(x_1,x_2,t)}{\partial x_2^2}=E\psi(x_1,x_2,t)$$

inside the well, with the wave function going to zero for x_1 or x_2 equal to $L/2$ or $-L/2$.

On looking at this equation, we see it is *the same* as theSchrodinger equation for a *single* electron in a *two* dimensional square well, and so can be solved in the same way, by separation of variables. For example, the wave function we plotted for the two dimensional rectangular well is in the square case:

$$\psi_{(2,3)}(x_1,x_2,t) = A\sin(2\pi x_1/L)\cos(3\pi x_2/L)e^{-Bt/k}$$

has the same energy

$$E = \frac{\hbar^2}{2m}\frac{\pi^2}{L^2}(2^2 + 3^2)$$

as the physically distinct wavefunction:

$$\psi_{(3,2)}(x_1, x_2, t) = A\sin(2\pi x_2 / L)\cos(3\pi x_1 / L)e^{-Bt/\hbar}$$

INTERPRETING THE WAVEFUNCTION

We have already discussed how the above wavefunctions are to be interpreted if they are regarded as two-dimensional wavefunctions for a single electron: $|\psi(x_1, x_2)|^2\, dx_1 dx_2$ is the probability of finding the electron in a small area $dx_1 dx_2$ at (x_1, x_2).

The natural extension of this interpretation to two electrons is to assume now $|\psi(x_1, x_2)|^2\, dx_1 dx_2$ is the *joint* probability of finding electron 1 in a small line length dx_1 at position x_1, and at the same time finding electron 2 in a small length dx_2 at x_2.

This, however, leads to a real problem. Consider the wavefunction

$$\psi_{(2,3)}(x_1, x_2, t) = A\sin(2\pi x_1 / L)\cos(3\pi x_2 / L)e^{-Bt/\hbar}$$

again. Let us now take specific points, $x_1 = 0$, $x_2 = L/4$. Then the probability of finding electron 1 in an infinitesimal interval at x_1 and electron 2 similarly at x_2 is zero, because y is zero at $x_1 = 0$. On the other hand, the probability of finding electron 2 at x_1 (= 0) and electron 1 at x_2 (= $L/4$) is *not* zero.

The problem is this: the electrons are *identical* (we assume their spins point the same way). We can't tell which is which, and nobody else can either. The indistinguishability of elementary particles is not like that of apparently identical macroscopic objects, where one could always place some tiny mark.

There is no way to mark an electron. This means, though, that the best we can do is to talk about the probability of finding one electron at x_1 and another at x_2, we *cannot specify which electron we find where.* Therefore, any alleged wave function that gives different probabilities for finding electron 1 at x_1, 2 at x_2 and finding 2 at x_1, 1 at x_2 is not physically meaningful.

That is to say, a wave function describing two identical particles must have a symmetric probability distribution. This is definitely not the case with our function, so although it is a solution to the two particleSchrodinger equation, it is not a physically meaningful wave function for two particles in a box.

$$|\psi(x_1, x_2)|^2 = |\psi(x_1, x_2)|^2.$$

In fact, this is not difficult to fix recall that the function $\psi_{(3,2)}(x_1, x_2)$ has the same energy, and in fact just corresponds to the two particles being switched around, that is to say $\psi_{(3,2)}(x_1, x_2), = \psi_{(2,3)}(x_1, x_2)$.

It follows that the symmetric function

$$\psi^s_{(2,3)}(x_1, x_2) = \frac{1}{\sqrt{2}}(\psi_{(2,3)}(x_1, x_2) + \psi_{(2,3)}(x_2, x_1))$$

and the antisymmetric function

$$\psi^s_{(2,3)}(x_1, x_2) = \frac{1}{\sqrt{2}}(\psi_{(2,3)}(x_1, x_2) - \psi_{(2,3)}(x_2, x_1))$$

are both solutions toSchrodinger 's equation for the energy E, and both satisfy the requirement $|\psi(x_1, x_2)|^2 = |\psi(x_1, x_2)|^2$ necessary for identical particles, so these are the appropriate candidate wave functions for the two particles in the one-dimensional box.

BOSONS, FERMIONS AND THE PAULI EXCLUSION PRINCIPLE

It turns out that both symmetric and antisymmetric wavefunctions arise in nature in describing identical particles. In fact, all elementary particles are either *fermions*, which have antisymmetric multiparticle wavefunctions, or *bosons*, which have symmetric wave functions. Electrons, protons and neutrons are fermions; photons, a-particles and helium atoms are bosons.

It is important to realise that this requirement of symmetry of the probability distribution, arising from the true indistinguishability of the particles, has a large effect on the probability distribution, and, furthermore, the effect is very different for fermions and bosons.

The simplest way to see this is just to plot $|\psi(x_1, x_2)|^2$ for the symmetric and for the antisymmetric wave functions, and compare them. The difference is dramatic. For the antisymmetric wave function, the particles are most likely to be found *far away* from each other. In fact, there is *zero* probability that they will be found at the same spot, because if $\psi(x_1, x_2) = -\psi(x_1, x_2)$, obviously $\psi(x, x) = 0$. A more general statement is that *two fermions cannot be in the same quantum state*, because if they were, the wave function would be of the symmetric form $f(x_1)f(x_2)$, and could not be antisymmetrized.

This is the Pauli Exclusion Principle it is the basis of the periodic table, and consequently of almost everything else.

We should perhaps emphasize that these wave functions were calculated with the electrostatic repulsion between the electrons switched off. That is *not* what is keeping the electrons apart, although it will increase the separation if it is included. We should also warn against a simple classical picture of the Pauli principle, the thought that two things can't be in the same place, after all, so perhaps it's no surprise two electrons can't be in the same state.

Two electrons *can* be in the same identical *space* wave function provided that their spins point in opposite ways. Furthermore, two bosons can be in the same state, and although that is perhaps reasonable sounding for photons, it is equally true for heavy atoms. In Bose-Einstein Condensation, a large number of atoms occupy the same quantum state. This happens in liquid Helium^4 below about two kelvin, it becomes a superfluid and flows without friction. BE condensation has also been achieved with laser cooled collections of large atoms in a trap.

ANGULAR MOMENTUM QUANTIZATION

THE STERN-GERLACH EXPERIMENT

We've established that for the hydrogen atom, the angular momentum of the electron's orbital motion has values $\sqrt{l(l+1)}\hbar$, where $l = 0, 1, 2, \ldots$, and the component of angular momentum in the z-direction is, where m takes integer values $-l, -l+1, \ldots, +l$. This means that if we measure the angle between the total angular momentum and the z-axis, there can only be $2l + 1$ possible answers, the total angular momentum cannot point in an arbitrary direction relative to the z-axis, odd though this conclusion seems. This is sometimes called "space quantization".

Is there any way we can actually *see* some effect of this directional quantization? The answer is yes because the electron moving around its orbit is a tiny loop of electric current, and, therefore, an electromagnet. So, if we switch on a magnetic field in the direction of the z-axis, the energy of the atom will depend on the degree of alignment of its magnetic moment with the external applied magnetic field. The magnetic field from a small current loop is like that from a small bar magnet aligned along the axis of the loop.

The simplest way to see how the potential energy of the little magnet depends on which way it's pointing relative to the field is to take a little bar with an N pole of strength $+p$ at one end, an S pole of strength $-p$ at the other end. Think of a compass needle, of length d, say.

The magnetic moment is defined as pole strength multiplied by distance between the poles, $\mu = pd$, and is considered to be a vector pointing along the axis of the magnet, from S to N. The potential energy of this little magnet in an external field H is -$\mu.H$, lowest when the magnet is fully aligned with the field. It is easy to check this: counting as zero potential energy the magnet at right angles with the field, the work needed to point it at an angle θ is $2.p.(d/2).\cos\theta$.

The magnetic moment of a current I going in a circle around an area A is just IA. The electron has charge e, and speed v, so goes around $v/2\pi r$ times per second. In other words, if you stand at one point in the orbit, the total charge passing you per second is $ev/2\pi r = I$. Hence the magnetic moment, usually denoted $\mu_L = IA$, is $\pi r^2.ev/2\pi r = rev/2$. The angular momentum is $L = mvr$, so

$$\mu_L = eL/2m.$$

Thus if the electron is in an $l = 1$ orbit, the current will generate a magnetic moment $e\hbar/2m$, which is 9.3×10^{-24} joules per tesla, or 5.8×10^{-5} eV per tesla. Note that this means in a one tesla field an atomic energy level will move ~10^{-4} eV, an easily detectable shift in spectral lines will result.

But there is a more direct way to see how the atoms are oriented, the Stern Gerlach apparatus. In this experiment, a beam of atoms is sent into a *nonuniform* magnetic field. This means the north and south poles of a small bar magnet would feel different strength forces, so there would be a net force on a small magnet, and hence on an atom. Furthermore, the direction of this force would depend on the orientation of the dipole.

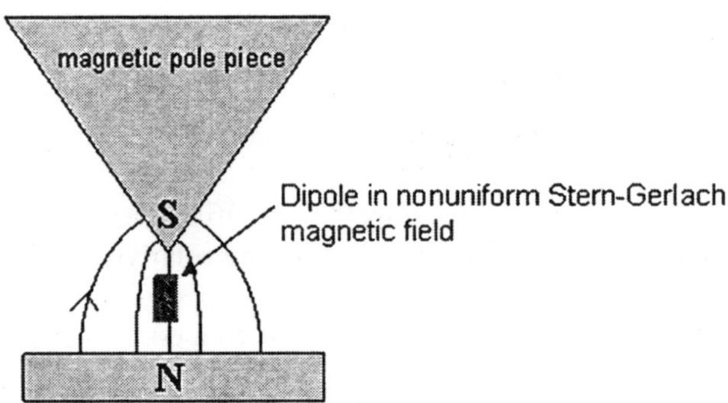

Suppose the nonuniform field is pointing upwards, and is stronger at the top. Then a small bar magnet oriented vertically with the north pole on top will be pushed upwards, because the north pole will be experiencing the stronger force. If the south pole is on top, the magnet will be pushed downwards. If the magnet is horizontal, there will be no net force (assuming magnetic field strength varies only negligibly in the horizontal direction).

Imagine, then, a stream of atoms with magnetic moments entering a region of magnetic field as described. Each atom will feel a vertical force depending on the orientation of its magnetic moment. If with no magnetic field present the stream of atoms formed a dot on a screen after passing through the apparatus, on switching on the field one would expect the dot to be stretched into a vertical line, if one assumed equal likelihood of all orientations of the magnetic moment.

However, the quantum theory predicts that this is not the case—we have argued that for $l = 1$, say, there are only three allowed orientations of the magnet (atom) relative to the field. Therefore, we would predict that three dots (or, more realistically, blobs) would appear on the screen, not a continuous line.

In fact, when the experiment was carried out, there was a

very surprising result. Perhaps the most dramatic form of the new result came later, in 1927, when *ground state $l = 0$ hydrogen atoms* were used (Phipps and Taylor, Phys Rev 29, 309). Such atoms have *no* orbital angular momentum, and therefore no orbital current, and were not expected to show magnetic effects.

Yet on going through the Stern-Gerlach apparatus, the beam of hydrogen atoms *split into two*! This was difficult to interpret, because the least allowed angular momentum, $l = 1$, would give three blobs, and $l = 0$ would give only one. You might expect a mixture to give one strong blob and two weak ones, but two equal blobs didn't seem possible, theoretically. Stern and Gerlach had themselves seen two blobs with silver atoms in 1922. We mention the hydrogen case first because it was by far the best understood atom (and still is!) so the need for new physics was clearest.

The solution to the problem was suggested by two graduate students, Goudsmit and Uhlenbeck. They suggested that the electron *itself* had a spin. That is to say, the electron both orbited the proton *and* spun on its own axis, just as the earth orbits the sun once a year and also spins on its own axis once a day. If the electron spin is assumed to be $\hbar/2$, and we assume as before that the z-component can only change by whole units of $\hbar$, then there are only *two* allowed values of the z-component $\pm\hbar/2$.

Of course, this is a hand waving argument the reason the z-component only changed by integers was that the wave function had to fit a whole number of wavelengths on going around the z-axis. But our wave function for spin one-half, if it is of the same form as those for angular momentum, must have a term $e^{i\varphi/2}$, and so is multiplied by -1 on rotating through 2π! (In fact, that the z-component can only change by whole units of $\hbar$ follows from very general properties of angular momentum.)

Further difficulties arose when people tried to construct models of how a spinning electron would have its own magnetic moment. It's not too difficult to see how this might occur—if the electron is a charged sphere, or has charge on its surface, then its rotation implies that this charge is going around in circles, little current loops, and so will give a magnetic field. The problem was, it was known that the electron was a very small object. It turned out that the equatorial speed of the electron would have to be greater than the speed of light for the magnetic moment to be of the observed strength. These difficulties in understanding the electron spin and magnetic moment were far from trivial, and in fact were not resolved until around 1930, by Dirac, who gave a fully relativistic treatment of the problem, which, remarkably, predicted the magnetic moment correctly and at the same time treated the electron as a point particle. There is no simple picture presenting this in classical or semiclassical terms, but Dirac's work is the basis of our modern understanding of particle physics. It is unfortunately beyond the scope of this course.

The bottom line, as far as we are concerned, is that assuming the electron has spin one-half and hence two possible spin orientations with respect to a given axis explains the observed Stern-Gerlach results, and also, more importantly, helps us construct the periodic table.

BUILDING THE PERIODIC TABLE

The *atomic number* (usually denoted *Z*) of an element denotes its place in the periodic table, so H has $Z = 1$; He, $Z = 2$; Li = 3, Be = 4, B = 5, C = 6, N = 7, O = 8, F = 9, Ne = 10, and so on. This number is equal to the number of protons in the nucleus, and also equal to the number of electrons orbiting around the nucleus, to preserve electrical neutrality.

To try to understand how the electrons orbit the nucleus, we need to make some simplifying assumptions. We are not going to be able to solveSchrodinger 's equation for even two electrons exactly, if we include their repulsion of each other. However, the presence of the other electrons is clearly important—their repulsion to some extent counteracts the attraction the nucleus has for a given electron.

For an electron imagined to be in some outer orbit, the electrons in closer to the nucleus orbits lower the effective nuclear charge. Thinking now about the force felt by one electron, a simple approximation is to imagine all the other electrons as changing the electrical attraction the one electron feels from the nucleus to a shielded attraction, so that the further it is away from the nucleus, the weaker an attractive charge it sees.

We then make the naove assumption that all the electrons

see the same potential, this shielded Coulomb potential, so we have Z electrons all in the same potential well, but we assume they are independent particles, in the sense that they do not repel each other, except to the extent already taken into account by changing to a shielded potential.

So the question is, what are the possible wave functions of Z independent electrons in this well? The crucial point is that although they do not interact with each other, they are identical, so the wave function must be antisymmetrized. This means that the electrons must be in different bound states in the well the Pauli exclusion principle.

But what do the bound state wave functions look like in this potential? Since the shielded Coulomb potential is still spherically symmetric, all our arguments about the θ,φ behaviour of the ordinary Coulomb potential apply equally to the shielded case, in particular the angular momentum has values $\sqrt{l(l+1)}\hbar$, where $l = 0, 1, 2, \ldots$, and the component of angular momentum in the z-direction is $m\hbar$, where m takes integer values $-l, -l+1, \ldots, +l$. Furthermore, each electron has spin $\hbar/2$, and there are two allowed values of the spin z-component, $\pm\hbar/2$. The radial wave functions R(r) are clearly somewhat different from those in the pure Coulomb case. The main difference is that states of different angular momentum which were degenerate in the Coulomb case are no longer the same energy in the shielded Coulomb case. If you examine wave functions corresponding to the same energy but different values of l, you will see that the higher the l-value, the smaller the wave function is near the nucleus. This means the higher l wave functions do not feel the powerful unshielded potential near the nucleus, and so are not as strongly bound as the lower l functions.

NOTATION

A standard notation is used by atomic physicists to describe these states. The different angular momenta are denoted by letters, s for $l = 0$, p for $l = 1$, d for $l = 2$, f for $l = 3$, g for $l = 4$ and then on alphabetically. The Principal Quantum Number n, such that for the hydrogen atom $E = -1/n^2$ in Rydberg units, is given as a number, so the lowest hydrogen atom state is written $1s$. The two $n = 2$

orbital states are $2s$ and $2p$, then come $3s$, $3p$ and $3d$ and so on. From the discussion immediately above, $2s$ and $2p$ have the same energy in the hydrogen atom, but for the shielded potential used to approximate for the presence of other electrons in bigger atoms $2s$ would be more tightly bound, and so at a lower energy, than $2p$.

FILLING AN ATOM WITH ELECTRONS

Let us now consider taking a bare nucleus, charge Z, and adding Z electrons to it one by one. From the Pauli Exclusion Principle, each electron must be in a different state. But remember that having a different spin counts as different (you could tell them apart) so we can put two electrons, with opposite spins, into each orbital state. Thus He has two electrons in the $1s$ state. Li must have two electrons in $1s$, and one electron in $2s$. This suggests a picture of one electron outside of a "closed shell" of two 1s electrons.

The next occurrence of a similar picture is Na, having $Z = 11$, which is chemically very similar to Li. This means that 10 electrons fill closed shells. We can understand this because 2 go into $1s$, 2 go into $2s$ and 6 fill $2p$. But notice by saying it takes 6 electrons to fill $3p$, we are saying there are three distinct $l = 1$ orbitals. In other words, *the chemical properties of the elements support and confirm the hypothesis of "space quantization"* that there are only three distinct $l = 1$ angular wave functions, those given by $m = 1$, 0 and -1.

Atoms interact chemically by sharing or partially transferring electrons. It's easier to transfer an electron that is loosely bound, and easier to accept one if there's a "hole" in a shell. Not surprisingly, atoms with filled shells only, like He and Ne, are chemically unreactive. The valency, roughly speaking, is the number of electrons available for transfer (so Li and Na have valency 1) or available sites for reception of electrons—fluorine has an outer shell with one vacancy, so a valency of 1. To some extent, valency can vary depending on the strength of attraction of other atoms in the chemical environment.

FILLING A BOX WITH ELECTRONS

When many Li atoms are put together to form a solid, it is

found that the loosely attached outer electrons leave their original atoms and wander freely throughout the metal. Their wave functions are well represented by standing plane waves in a box (let's take a cube of metal, of side L).

Each such plane wave state in the box can be represented by three numbers n_x, n_y and n_z, representing the number of nodes of the standing wave in the x, y and z directions respectively. Extending slightly our analysis of an electron in a two-dimensional box, the energy of such a state will be

$$E = \frac{\hbar^2}{2m}\frac{\pi^2}{L^2}(n_x^2 + n_y^2 + n_z^2).$$

Thus if we imagine pouring electrons into an empty lattice of Li atoms each with one electron missing(not a physically realistic procedure!) two electrons (opposite spins) will go into each state, first (0, 0, 0) then (1, 0, 0) or equally (0, 1, 0) etc., and from the form of the energy we can see that in (n_x, n_y, n_z) space, the electrons will fill up all the positive integer points within a sphere up to some maximum energy determined by how many electrons we put in. Notice that since the n's are all positive integers, the filled space is only the one-eighth of the sphere's volume corresponding to $n_x > 0$, $n_y > 0$ and $n_z > 0$ for the sphere centered at the origin.

Physicists sometimes formulate this filling of electron states slightly differently, by imposing periodic boundary conditions on a piece of metal, like replacing a finite line by a ring. This is not easy to do in three dimensions, but is convenient to talk about. The advantage is that instead of standing waves, all the electrons have definite momenta.

The allowed momenta form a grid in "momentum space" a lot like the allowed integers in the standing waves above. In fact it turns out that there are the same number of allowed momenta up to a certain energy as there are allowed standing wave states. The difference is that in momentum space, if momentum k is allowed, so is $-k$, and in the ground state momentum states are filled up to a spherical surface, called the "Fermi Surface"—an energy equipotential at the "Fermi energy".

Typical Fermi energies are of the order of electron volts. By spreading out through the metal in this way the electrons attain an overall lower energy state than if each stayed with its own atom. This is why the solid is stable.

On applying heat to the electrons, even 1000K is only 0.1eV, so only those near the surface of the filled sphere are free to move, because of the exclusion principle, the others are locked in.

This means the heat capacity of the electrons is much less than $(3/2)kT$ per particle, as would be predicted classically. This was another long standing classical puzzle solved by the advent of quantum mechanics.

Chapter 6

Kinetic Theory of Gases

BERNOULLI'S PICTURE

Daniel Bernoulli, in 1738, was the first to understand air pressure from a molecular point of view. He drew a picture of a vertical cylinder, closed at the bottom, with a piston at the top, the piston having a weight on it, both piston and weight being supported by the air pressure inside the cylinder. He described what went on inside the cylinder as follows: "let the cavity contain very minute corpuscles, which are driven hither and thither with a very rapid motion; so that these corpuscles, when they strike against the piston and sustain it by their repeated impacts, form an elastic fluid which will expand of itself if the weight is removed or diminished..."

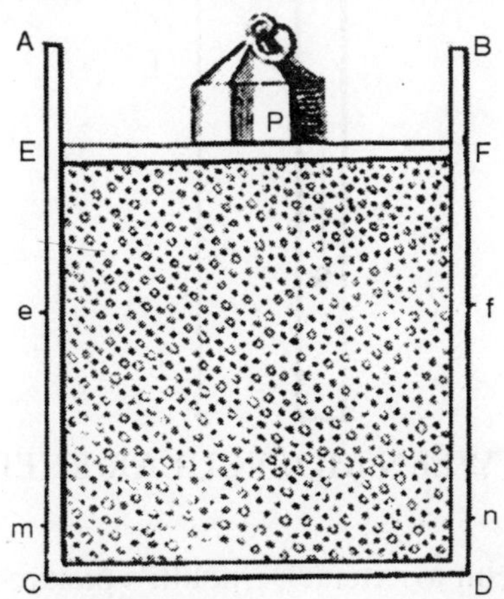

(An applet is available here.) Sad to report, his insight,

although essentially correct, was not widely accepted. Most scientists believed that the molecules in a gas stayed more or less in place, repelling each other from a distance, held somehow in the ether. Newton had shown that $PV =$ constant followed if the repulsion were inverse-square. In fact, in the 1820's an Englishman, John Herapath, derived the relationship between pressure and molecular speed given below, and tried to get it published by the Royal Society.

It was rejected by the president, Humphry Davy, who pointed out that equating temperature with motion, as Herapath did, implied that there would be an absolute zero of temperature, an idea Davy was reluctant to accept. And it should be added that no-one had the slightest idea how big atoms and molecules were, although Avogadro had conjectured that equal volumes of different gases at the same temperature and pressure contained equal numbers of molecules his famous number neither he nor anyone else knew what that number was, only that it was pretty big.

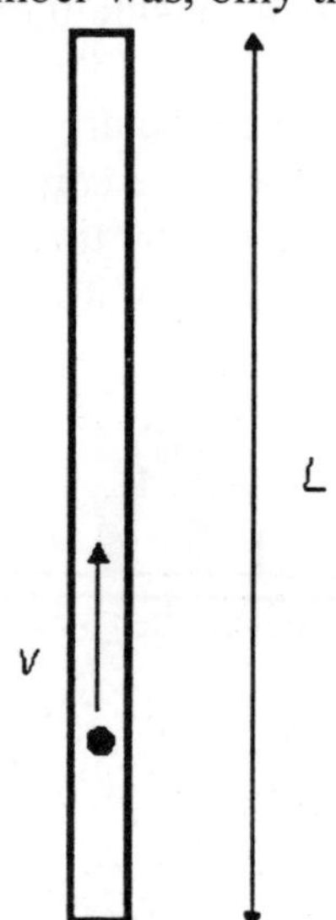

THE LINK BETWEEN MOLECULAR ENERGY AND PRESSURE

It is not difficult to extend Bernoulli's picture to a *quantitative* description, relating the gas pressure to the molecular velocities.

Let us consider a single perfectly elastic particle, of mass m, bouncing rapidly back and forth at speed v inside a narrow cylinder of length L with a piston at one end, so all motion is along the same line. What is the force on the piston? Obviously, the piston doesn't feel a smooth continuous force, but a series of equally spaced impacts. However, if the piston is much heavier than the particle, this will have the same effect as a smooth force over times long compared with the interval between impacts. So what is the value of the equivalent smooth force?

Using Newton's law in the form *force = rate of change of momentum*, we see that the particle's momentum changes by $2mv$ each time it hits the piston. The time between hits is $2L/v$, so the frequency of hits is $v/2L$ per second. This means that if there were no balancing force, by conservation of momentum the particle would cause the momentum of the piston to change by $2mv'v/2L$ units in each second. This is the rate of change of momentum, and so must be equal to the balancing force, which is therefore $F = mv^2/L$.

We now generalize to the case of many particles bouncing around inside a rectangular box, of length L in the x-direction (which is along an edge of the box). The total force on the side of area A perpendicular to the x-direction is just a sum of single particle terms, the relevant velocity being the component of the velocity in the x-direction. The *pressure* is just the force per unit area, $P = F/A$. Of course, we don't know what the velocities of the particles are in an actual gas, but it turns out that we don't need the details. If we sum N contributions, one from each particle in the box, each contribution proportional to v_x^2 for that particle, the sum just gives us N times the *average* value of v_x^2. That is to say,

$$P = F/A = Nm\overline{v_x^2} / LA = Nm\overline{v_x^2} / V$$

where there are N particles in a box of volume V. Next we note that the particles are equally likely to be moving in any direction, so the average value of v_x^2 must be the same as that of v_y^2 or v_z^2, and since $v^2 = v_x^2 + v_y^2 + v_z^2$, it follows that

$$P = Nm\overline{v_x^2} / 3V\,.$$

This is a surprisingly simple result! *The macroscopic pressure of a gas relates directly to the average kinetic energy per molecule.* Of course, in the above we have not thought about possible complications caused by interactions between particles, but in fact for gases like air at room temperature these interactions are very small. Furthermore, it is well established experimentally that most gases satisfy the Gas Law over a wide temperature range:

$$PV = nRT$$

for n moles of gas, that is, $n = N/N_A$, with N_A Avogadro's number and R the gas constant. Introducing Boltzmann's constant $k = R/N_A$, it is easy to check from our result for the pressure and the ideal gas law that *the* average molecular kinetic energy is proportional to the absolute temperature,

$$\overline{E_k} = \overline{\frac{1}{2}mv^2} = \frac{3}{2}kT.$$

Boltzmann's constant $k = 1.38.10^{-23}$ joules/K.

MAXWELL FINDS THE VELOCITY DISTRIBUTION

By the 1850's, various difficulties with the existing theories of heat, such as the caloric theory, caused some rethinking, and people took another look at the kinetic theory of Bernoulli, but little real progress was made until Maxwell attacked the problem in 1859. Maxwell worked with Bernoulli's picture, that the atoms or molecules in a gas were perfectly elastic particles, obeying Newton's laws, bouncing off each other (and the sides of the container) with straight-line trajectories in between collisions.

(Actually, there is some inelasticity in the collisions with the sides the bouncing molecule can excite or deexcite vibrations in the wall, this is how the gas and container come to thermal equilibrium.) Maxwell realized that it was completely hopeless to try to analyse this system using Newton's laws, even though it could be done in principle, there were far too many variables to begin writing down equations.

On the other hand, a completely detailed description of how each molecule moved was not really needed anyway. What *was* needed was some understanding of how this microscopic picture connected with the macroscopic properties, which represented averages over huge numbers of molecules.

The relevant microscopic information is not knowledge of the position and velocity of every molecule at every instant of time, but just the distribution function, that is to say, what percentage of the molecules are in a certain part of the container, and what percentage have velocities within a certain range, at each instant of time. For a gas in thermal equilibrium, the distribution function is independent of time. Ignoring tiny corrections for gravity, the gas will be distributed uniformly in the container, so the only unknown is the *velocity* distribution function.

VELOCITY SPACE

What does a velocity distribution function look like? Suppose at some instant in time one particular molecule has velocity $\vec{v} = (v_x, v_y, v_z)$. We can record this information by constructing a three-dimensional *velocity space*, with axes v_x, v_y, v_z, and putting in a point P_1 representing the molecule's velocity (the red arrow is of course $\vec{v}$):

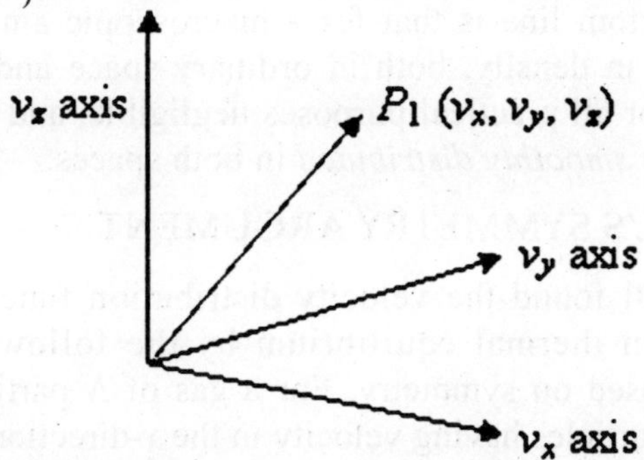

Fig. Point P_1 represents the location of one molecule in velocity space

Now imagine that at that instant we could measure the velocities of *all* the molecules in a container, and put points P_2, P_3, P_4, ... P_N in the velocity space. Since N is of order 10^{21} for 100 ccs of gas, this is not very practical! But we can imagine what the result would be: a cloud of points in velocity space, equally spread in all directions (there's no reason molecules would prefer to be moving in the x-direction, say, rather than the y-direction) and thinning out on going away from the origin towards higher and higher velocities.

Now, if we could keep monitoring the situation as time passes individual points would move around, as molecules bounced off the walls, or each other, so you might think the cloud would shift around a bit. But there's a *vast number* of molecules in any realistic macroscopic situation, and for any reasonably sized container it's safe to assume that the number of molecules in any small region of velocity space remains pretty much constant.

Obviously, this cannot be true for a region of velocity space so tiny that it only contains one or two molecules on average. But it can be shown statistically that if there are N molecules in a particular small volume of velocity space, the fluctuation of the number with time is of order $\sqrt{N}$, so a region containing a million molecules will vary in numbers by about one part in a thousand, a trillion molecule region by one part in a million. Since 100 ccs of air contains of order 10^{21} molecules, we can in practice divide the region of velocity space occupied by the gas into a billion cells, and *still* have variation in each cell of order one part in a million!

The bottom line is that for a macroscopic amount of gas, fluctuations in density, both in ordinary space and in velocity space, are for all practical purposes negligible, and we can take the gas to be *smoothly distributed* in both spaces.

MAXWELL'S SYMMETRY ARGUMENT

Maxwell found the velocity distribution function for gas molecules in thermal equilibrium by the following elegant argument based on symmetry. For a gas of N particles, let the number of particles having velocity in the x-direction between v_x and $v_x + dv_x$ be $Nf_1(vx)dv_x$. In other words, is the fraction of all the particles having x-direction velocity lying in the interval between v_x and $v_x + dv_x$. (I've written f_1 instead of f to help remember this function refers to only one component of the velocity vector.) If we add the fractions for all possible values of v_x, the result must of course be 1:

$$\int_{-\infty}^{\infty} f_1(v_x)dv_x = 1.$$

But there's nothing special about the x-direction—for gas

molecules in a container, at least away from the walls, all directions look the same, so the same function f will give the probability distributions in the other directions too. It follows immediately that the probability for the velocity to lie between v_x and $v_x + dv_x$, v_y and $v_y + dv_y$, and v_z and $v_z + dv_z$ must be:

$$Nf_1(v_x)dv_x f_1(v_y)dv_y f_1(v_z)dv_z =$$
$$Nf_1(v_x)dv_x f_1(v_y)f_1(v_z)dv_x dv_y dv_z$$

Note that this distribution function, when integrated over all possible values of the three components of velocity, gives the total number of particles to be N, as it should (since integrating over each $f_1(v)dv$ gives unity). Next comes the clever part since any direction is as good as any other direction, the distribution function must depend *only* on the total speed of the particle, *not* on the separate velocity components. Therefore, Maxwell argued, it *must* be that:

$$f_1(v_x)f_1(v_y)f_1(v_z) = F(v_x^2 + v_y^2 + v_z^2)$$

where F is another unknown function. However, it is apparent that the product of the functions on the left is reflected in the sum of variables on the right. It will only come out that way if the variables appear in an exponent in the functions on the left. In fact, it is easy to check that this equation is solved by a function of the form:

$$f_1(v_x) = Ae^{-Bv_x^2}.$$

This curve is called a *Gaussian*: it's centered at the origin, and falls off *very* rapidly as v_x increases. Taking $A = B = 1$ just to see the shape, we find:

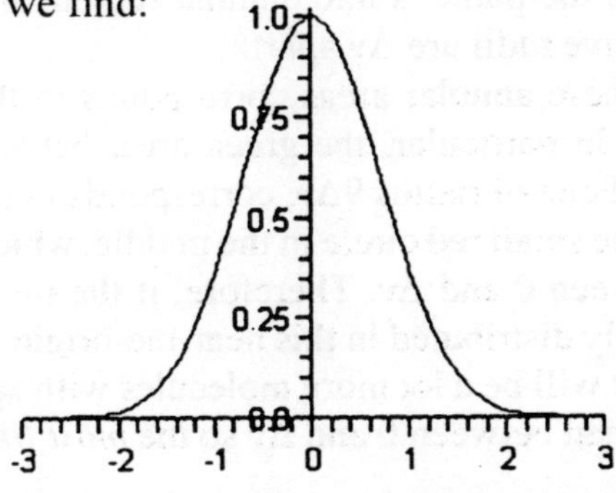

At this point, A and B are arbitrary constants—we shall

eventually find their values for an actual sample of gas at a given temperature. Notice that (following Maxwell) we have put a minus sign in the exponent because there must eventually be fewer and fewer particles on going to higher speeds, certainly not a diverging number.

Multiplying together the probability distributions for the three directions gives the distribution in terms of particle speed v, where $v^2 = v_x^2 + v_y^2 + v_z^2$. Since all velocity directions are equally likely, it is clear that the natural distribution function is that giving the number of particles having *speed* between v and $v + dv$.

From the graph above, it is clear that the most likely value of v_x is zero. If the gas molecules were restricted to one dimension, just moving back and forth on a line, then the most likely value of their *speed* would *also* be zero. However, for gas molecules free to move in two or three dimensions, the most likely value of the speed is *not* zero.

It's easiest to see this in a two-dimensional example. Suppose we plot the points *P* representing the velocities of molecules in a region near the origin, so the density of points doesn't vary much over the extent of our plot (we're staying near the top of the peak in the one-dimensional curve shown above).

Now divide the two-dimensional space into regions corresponding to equal increments in speed:

$$0 \text{ to } \Delta v, \Delta v \text{ to } 2\Delta v, 2\Delta v \text{ to } 3\Delta v, \ldots$$

In the two-dimensional space $v = \sqrt{v_x^2 + v_y^2}$, is a circle, so this division of the plane is into annular regions between circles whose successive radii are Δv apart:

Each of these annular areas corresponds to the same speed increment Δv. In particular, the green area, between a circle of radius $8\Delta v$ and one of radius $9\Delta v$, corresponds to the same *speed increment* as the small red circle in the middle, which corresponds to speeds between 0 and Δv. Therefore, if the molecular speeds are pretty evenly distributed in this near-the-origin area of the (v_x, v_y) plane, there will be a lot more molecules with speeds between $8\Delta v$ and $9\Delta v$ than between 0 and Δv so the *most likely speed* will *not* be zero.

To find out what it actually is, we have to put this area

argument together with the Gaussian fall off in density on going far from the origin. We'll discuss this shortly.

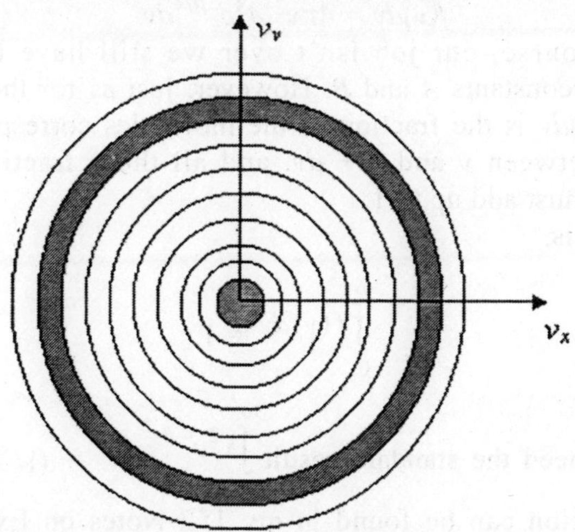

Fig. Constant speed circles in a two-dimensional example.

The same argument works in three dimensions it's just a little more difficult to visualize. Instead of concentric circles, we have concentric spheres. All points lying on a spherical surface centered at the origin correspond to the same speed.

Let us now figure out the distribution of particles as a function of speed. The distribution in the three-dimensional space (v_x, v_y, v_z) is from Maxwell's analysis

# of particles in small box $dv_x dv_y dv_z$

$$= Nf_1(v_x)f_1(v_y)f_1(v_z)dv_x dv_y dv_z$$

$$= NA^3 e^{-Bv2}\, dv_x dv_y dv_z$$

To translate this to the number of particles having *speed* between v and $v + dv$ we need to figure out how many of those little $dv_x dv_y dv_z$ boxes there are corresponding to speeds between v and $v + dv$. In other words, what is the *volume* of velocity space between the two neighboring spheres, both centered at the origin, the inner one with radius v, the outer one infinitesimally bigger, with radius $v + dv$? Since dv is so tiny, this volume is just the area of the sphere multiplied by dv: that is, $4\pi v^2 + dv$.

Finally, then, the probability distribution as a function of speed is:

$$f(v)dv = 4\pi v^2 A^3 e^{-Bt^2} dv.$$

Of course, our job isn't over we still have these two unknown constants A and B. However, just as for the function $f_1(v_x)$, $f(v)dv$ is the fraction of the molecules corresponding to speeds between v and $v + dv$, and all these fractions taken together must add up to 1.

That is,

$$\int_0^\infty f(v)\,dv = 1.$$

We need the standard result $\int_0^\infty x^2 e^{-Bx^2}\,dx = (1/4B)\sqrt{\pi/B}$ (a derivation can be found in my 152 Notes on Exponential Integrals), and find:

$$4\pi A^3 \frac{1}{4B}\sqrt{\frac{\pi}{B}} = 1.$$

This means that there is really only one arbitrary variable left: if we can find B, this equation gives us A: that is, $4\pi A^3 = \frac{4}{\sqrt{\pi}} B^{3/2}$, and $4\pi A^3$ is what appears in $f(v)$.

Looking at $f(v)$, we notice that B is a measure of how far the distribution spreads from the origin: if B is small, the distribution drops off more slowly the average particle is more energetic. Recall now that the average kinetic energy of the particles is related to the temperature by $\overline{\frac{1}{2}mv^2} = \frac{3}{2}kT$. This means that B is related to the inverse temperature.

In fact, since $f(v)dv$ is the fraction of particles in the interval dv at v, and those particles have kinetic energy $\frac{1}{2}mv^2$, we can use the probability distribution to find the average kinetic energy per particle:

$$\overline{\frac{1}{2}mv^2} = \int_0^\infty \frac{1}{2}mv^2 f(v)dv.$$

To do this integral we need another standard result:

$\int_0^\infty x^2 e^{-Bx^2}\, dx = (3/8B^2)\sqrt{\pi / B}$. We find:

$$\overline{\frac{1}{2}mv^2} = \frac{3m}{4B}.$$

Substituting the value for the average kinetic energy in terms of the temperature of the gas,

$$\overline{\frac{1}{2}mv^2} = \frac{3}{2}kT$$

gives $B = m/2kT$, so $4\pi A^2 = \frac{4}{\sqrt{\pi}} B^{3/2} = 4\pi \left(\frac{m}{2\pi kT}\right)^{3/2}$.

This means the distribution function

$$f(v) = 4\pi \left(\frac{m}{2\pi kT}\right)^{3/2} v^2 e^{-mv^2/2kT}$$

$$= 4\pi \left(\frac{m}{2\pi kT}\right)^{3/2} v^2 e^{-E/kT}$$

where E is the kinetic energy of the molecule.

Note that this function increases *parabolically* from zero for low speeds, then curves round to reach a maximum and finally decreases exponentially. As the temperature increases, the position of the maximum shifts to the right. *The total area under the curve is always one*, by definition. For air molecules (say, nitrogen) at room temperature the curve is the blue one below. The red one is for an absolute temperature down by a factor of two:

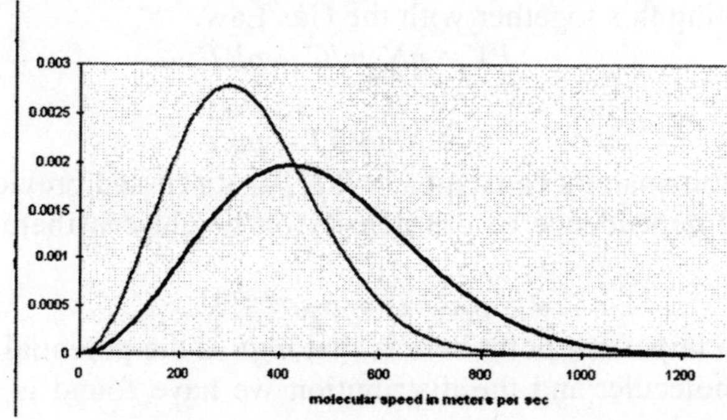

WHAT ABOUT POTENTIAL ENERGY

Maxwell's analysis solves the problem of finding the statistical velocity distribution of molecules of an ideal gas in a box at a definite temperature *T*: the relative probability of a molecule having velocity $\vec{v}$ is proportional to $e^{-mv^2/2T} = e^{-E/kT}$. The *position* distribution is taken to be uniform: the molecules are assumed to be equally likely to be anywhere in the box.

But how is this distribution affected if in fact there is some kind of potential pulling the molecules to one end of the box? In fact, we've already solved this problem, in the discussion earlier on the isothermal atmosphere.

Consider a really big box, kilometers high, so air will be significantly denser towards the bottom. Assume the temperature is uniform throughout. We found under these conditions that with Boyles Law expressed in the form

$$\rho = CP$$

the atmospheric density varied with height as

$$P = P_0 e^{-Cgh}, \text{ or equivalenty } \rho = \rho_0 e^{-Cgh}.$$

Now we know that Boyle's Law is just the fixed temperature version of the Gas Law $PV = nRT$, and the density

$$\rho = \text{mass/volume} = Nm/V$$

with N the total number of molecules and m the molecular mass,

$$CP = \rho = Nm/V$$

Rearranging,

$$PV = Nm/C = nN_A m/C,$$

for n moles of gas, each mole containing Avogadro's number N_A molecules.

Putting this together with the Gas Law,

$$PV = nN_A m/C = nRT,$$

so

$$C = N_A m/RT = mkT$$

where Boltzmann's constant $k = R/N_A$ as discussed previously.

The dependence of gas density on height can therefore be written

$$\rho = \rho_0 e^{-Cgh} = \rho_0 e^{-mgh/kT}.$$

The important point here is that *mgh* is the potential energy of the molecule, and the distribution we have found is exactly parallel to Maxwell's velocity distribution, the potential energy now playing the role that kinetic energy played in that case.

We're now ready to put together Maxwell's velocity distribution with this height distribution, to find out how the molecules are distributed in the atmosphere, *both* in velocity space *and* in ordinary space. In other words, in a *six*-dimensional space!

Our result is:

$$f(x, y, z, v_x, v_y, v_z) = f(h, v) \propto e^{-mv^2/2kT}$$
$$= e^{-(1/2)mv2+\text{mgh})/kT} = e^{-E/kT}.$$

That is, the probability of a molecule having total energy E is proportional to $e^{-E/kT}$.

This is the Boltzmann, or Maxwell-Boltzmann, distribution. It turns out to be correct for any type of potential energy, including that arising from forces between the molecules themselves.

DEGREES OF FREEDOM AND EQUIPARTITION OF ENERGY

By a "degree of freedom" we mean a way in which a molecule is free to move, and thus have energy—in this case, just the *x*, *y*, and *z* directions. Boltzmann reformulated Maxwell's analysis in terms of degrees of freedom, stating that there was an average energy ½kT in each degree of freedom, to give total average kinetic energy 3.½kT, so the specific heat per molecule is presumable 1.5*k*, and given that $k = R/N_A$, the specific heat per mole comes out at 1.5*R*. In fact, this is experimentally confirmed for monatomic gases. However, it is found that diatomic gases can have specific heats of 2.5*R* and even 3.5*R*.

This is not difficult to understand—these molecules have more degrees of freedom. A dumbbell molecule can rotate about two directions perpendicular to its axis. A diatomic molecule could also vibrate. Such a simple harmonic oscillator motion has both kinetic and potential energy, and it turns out to have total energy kT in thermal equilibrium.

Thus, reasonable explanations for the specific heats of various gases can be concocted by assuming a contribution ½*k* from each degree of freedom. But there are problems. Why shouldn't the dumbbell rotate about its axis? Why do monatomic atoms not rotate at all? Even more ominously, the specific heat of hydrogen, $2.5R$ at room temperature, drops to $1.5R$ at lower temperatures. These problems were not resolved until the advent of quantum mechanics.

BROWNIAN MOTION

One of the most convincing demonstrations that gases really *are* made up of fast moving molecules is Brownian motion, the observed constant jiggling around of tiny particles, such as fragments of ash in smoke. This motion was first noticed by a Scottish botanist, who initially assumed he was looking at living creatures, but then found the same motion in what he knew to be particles of inorganic material. Einstein showed how to use Brownian motion to estimate the size of atoms.

BLACK BODY RADIATION

HEATED BODIES RADIATE

We shall now turn to another puzzle confronting physicists at the turn of the century : just how do heated bodies radiate? There was a general understanding of the mechanism involved—heat was known to cause the molecules and atoms of a solid to vibrate, and the molecules and atoms were themselves complicated patterns of electrical charges. From the experiments of Hertz and others, Maxwell's predictions that oscillating charges emitted electromagnetic radiation had been confirmed, at least for simple antennas. It was known from Maxwell's equations that this radiation traveled at the speed of light and from this it was realized that light itself, and the closely related infrared heat radiation, were actually electromagnetic waves. The picture, then, was that when a body was heated, the consequent vibrations on a molecular and atomic scale inevitably induced charge oscillations.

Assuming then that Maxwell's theory of electromagnetic radiation, which worked so well in the macroscopic world, was also valid at the molecular level, these oscillating charges would radiate, presumably giving off the heat and light observed.

HOW IS RADIATION ABSORBED?

What is meant by the phrase "black body" radiation? The

point is that the radiation from a heated body depends to some extent on the body being heated. To see this most easily, let's back up momentarily and consider how different materials *absorb* radiation. Some, like glass, seem to absorb light hardly at all the light goes right through. For a shiny metallic surface, the light isn't absorbed either, it gets reflected. For a black material like soot, light and heat are almost completely absorbed, and the material gets warm.

How can we understand these different behaviors in terms of light as an electromagnetic wave interacting with charges in the material, causing these charges to oscillate and absorb energy from the radiation? In the case of glass, evidently this doesn't happen, at least not much. Why not? A full understanding of why needs quantum mechanics, but the general idea is as follows: there are charges electrons in glass that are able to oscillate in response to an applied external oscillating electric field, *but* these charges are tightly bound to atoms, and can only oscillate at certain frequencies. (For quantum experts, these charge oscillations take place as an electron moves from one orbit to another.

Of course, that was not understood in the 1890's, the time of the first precision work on black body radiation.) It happens that for ordinary glass *none of these frequencies corresponds to visible light*, so there is *no* resonance with a light wave, and hence little energy absorbed. That's why glass is perfect for windows! Duh. But glass *is* opaque at some frequencies *outside* the visible range (in general, both in the infrared and the ultraviolet). These are the frequencies at which the electrical charge distributions in the atoms or bonds can naturally oscillate.

How can we understand the *reflection* of light by a *metal* surface? A piece of metal has electrons free to move through the entire solid. This is what makes a metal a metal: it conducts both electricity and heat easily, both are actually carried by currents of these freely moving electrons. (Well, a little of the heat is carried by vibrations.)

But metals are recognizable because they're shiny—why's that? Again, it's those free electrons: they're driven into large (relative to the atoms) oscillations by the electrical field of the incoming light wave, and this induced oscillating current radiates electromagnetically, just like a current in a transmitting antenna. This radiation *is* the reflected light. For a shiny metal surface, little of the incoming radiant energy is absorbed as heat, it's just reradiated, that is, reflected.

Now let's consider a substance that *absorbs* light: no transmission and no reflection. We come very close to perfect absorption with soot. Like a metal, it will conduct an electric current, but nowhere near as efficiently. There *are* unattached electrons, which can move through the whole solid, but they constantly bump into things—they have a short mean free path.

When they bump, they cause vibration, like balls hitting bumpers in a pinball machine, so they give up kinetic energy into heat. Although the electrons in soot have a short mean free path compared to those in a good metal, they move very freely compared with electrons bound to atoms (as in glass), so they can accelerate and pick up energy from the electric field in the light wave. They are therefore very effective intermediaries in transferring energy from the light wave into heat.

RELATING ABSORPTION AND EMISSION

Having seen how soot can absorb radiation and transfer the energy into heat, what about the reverse? Why does it radiate when heated? The pinball machine analogy is still good: imagine now a pinball machine where the barriers, etc., vibrate vigorously because they are being fed energy.

The balls (the electrons) bouncing off them will be suddenly accelerated at each collision, and these accelerating charges emit electromagnetic waves. On the other hand, the electrons in a *metal* have very long mean free paths, the lattice vibrations affect them much less, so they are less effective in gathering and radiating away heat energy. It is evident from considerations like this that good absorbers of radiation are also good emitters.

In fact, we can be much more precise: a body emits radiation at a given temperature and frequency *exactly* as well as it absorbs the same radiation. This was proved by Kirchhoff: the essential point is that if we suppose a particular body can absorb better than it emits, then in a room full of objects all at the same temperature, it will absorb radiation from the other bodies better than it radiates energy back to them. This means it will get hotter, and the rest of the room will grow colder, contradicting the second law of thermodynamics. (We could use such a body to construct a heat engine extracting work as the room grows colder and colder!)

But a metal glows when it's heated up enough: why is that? As the temperature is raised, the lattice of atoms vibrates more and more, these vibrations scatter and accelerate the electrons. Even glass glows at high enough temperatures, as the electrons are loosened and vibrate.

THE "BLACK BODY" SPECTRUM: A HOLE IN THE OVEN

Any body at any temperature above absolute zero will radiate to some extent, the intensity and frequency distribution of the radiation depending on the detailed structure of the body. To begin analyzing heat radiation, we need to be specific about the body doing the radiating: the simplest possible case is an idealized body which is a perfect absorber, and therefore also (from the above argument) a perfect emitter. For obvious reasons, this is called a "black body".

But we need to check our ideas experimentally: so how do we construct a perfect absorber? OK, nothing's perfect, but in 1859 Kirchhoff had a good idea: a small hole in the side of a large box is an excellent absorber, since any radiation that goes through the hole bounces around inside, a lot getting absorbed on each bounce, and has little chance of ever getting out again.

So, we can do this *in reverse*: have an oven with a tiny hole in the side, and presumably the radiation coming out the hole is as good a representation of a perfect emitter as we're going to find. Kirchhoff challenged theorists and experimentalists to figure out and measure (respectively) the energy/frequency curve for this "cavity radiation", as he called it (in German, of course: hohlraumstrahlung, where hohlraum means hollow room or cavity, strahlung is radiation).

WHAT WAS OBSERVED: TWO LAWS

The first quantitative conjecture based on experimental

observation of hole radiation was: Stefan's Law : the *total* power P radiated from one square meter of black surface at temperature T goes as the *fourth power* of the absolute temperature:

$$P = \sigma T^4, \sigma = 5.67 \times 10^{-8} \text{ watts/sq.m/K}^4.$$

Five years later, in 1884, Boltzmann derived this T^4 behaviour from theory: he applied classical thermodynamic reasoning to a box filled with electromagnetic radiation, using Maxwell's equations to relate pressure to energy density.

(The tiny amount of energy coming out of the hole would of course have the same temperature dependence as the radiation intensity inside.)

Exercise: The sun's surface temperature is 5700K. How much power is radiated by one square meter of the sun's surface? Given that the distance to earth is about 2,000 sun radii, what is the maximum power possible from a one square kilometre solar energy installation?

Another important finding was Wien's Displacement Law: As the oven temperature varies, so does the frequency at which the emitted radiation is most intense. In fact, that frequency is directly proportional to the absolute temperature:

$$f_{max} = \propto T.$$

(Wien himself deduced this law theoretically in 1893, following Boltzmann's thermodynamic reasoning. It had previously been observed, at least semi-quantitatively, by an American astronomer, Langley.)

In fact, this upward shift in f_{max} with T is familiar to everyone when an iron is heated in a fire, the first visible radiation (at around 900K) is deep red, the lowest frequency visible light. Further increase in T causes the colour to change to orange then yellow, and finally blue at very high temperatures (10,000K or more) for which the peak in radiation intensity has moved beyond the visible into the ultraviolet.

This shift in the frequency at which radiant power is a maximum is very important for harnessing solar energy, such as in a greenhouse.

The glass must allow the solar radiation in, but not let the heat radiation out. This is feasible because the two radiations are in very different frequency ranges 5700K and, say, 300K and there are materials transparent to light but opaque to infrared radiation. Greenhouses only work because f_{max} varies with temperature.

WHAT WAS OBSERVED: THE COMPLETE PICTURE

By the 1890's, experimental techniques had improved sufficiently that it was possible to make fairly precise measurements of the energy distribution in this cavity radiation, or as we shall call it black body radiation. In 1895, at the University of Berlin, Wien and Lummer punched a small hole in the side of an otherwise completely closed oven, and began to measure the radiation coming out.

The beam coming out of the hole was passed through a diffraction grating, which sent the different wavelengths/ frequencies in different directions, all towards a screen.

A detector was moved up and down along the screen to find how much radiant energy was being emitted in each frequency range. (This is a theorist's model of the experiment actual experimental arrangements were much more sophisticated. For example, to make the difficult infrared measurements higher frequency waves were eliminated by multiple reflections from quartz and other crystals.)

They found a radiation intensity/frequency curve close to this (correct one):

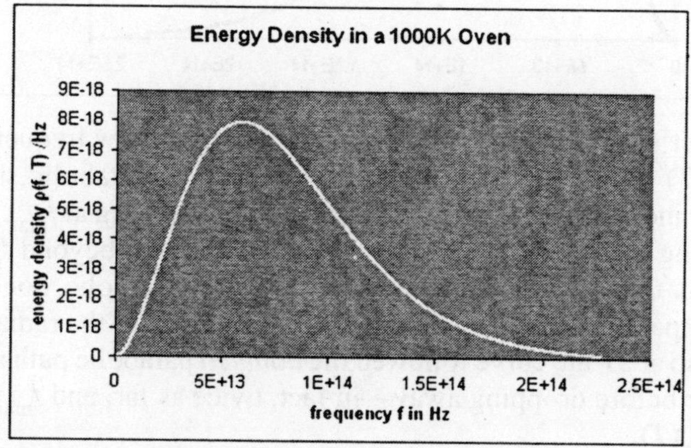

The visible spectrum begins at around 4.3×10^{14} Hz, so this oven glows deep red.

One minor point: this plot is the energy density *inside* the oven, which we denote by $\rho(f, T)$, meaning that at temperature T, the energy in Joules/m^3 in the frequency interval $f, f + \Delta f$ is $\rho(f, T)\Delta f$.

To find the power pumped out of the hole, bear in mind that the radiation inside the oven has waves equally going both ways—so only half of them will come out through the hole. Also, if the hole has area A, waves coming from the inside at an angle will see a smaller target area. The result of these two effects is that the radiation power from hole area $A = \frac{1}{4}Ac\,\rho(f, T)$.

They were also able to confirm both Stefan's Law $P = \sigma T^4$ and Wien's Displacement Law by measuring the black body curves at different temperatures, for example:

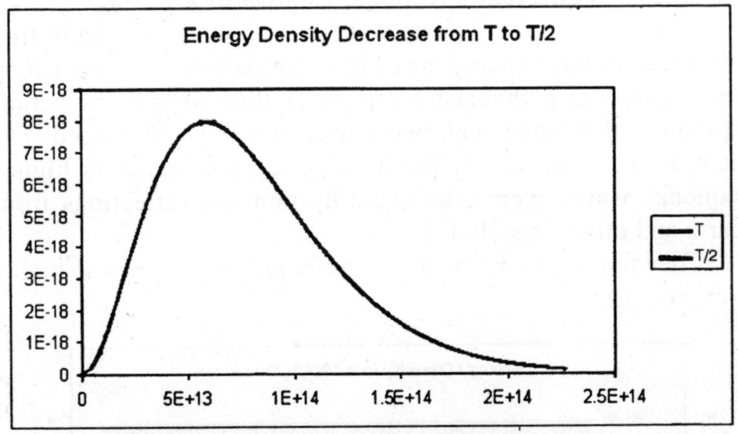

Let's look at these curves in more detail: for low frequencies f, $\rho(f, T)$ was found to be proportional to f^2, a parabolic shape, but for increasing f it fell below the parabola, peaking at f_{max}, then dropping quite rapidly towards zero as f increased beyond f_{max}.

For those low frequencies where $\rho(f, T)$ is parabolic, doubling the temperature was found to double the intensity of the radiation. But also at $2T$ the curve followed the *doubled* parabolic path much further before dropping away—in fact, twice as far, and $f_{max}(2T) = 2f_{max}(T)$.

The curve $\rho(f, 2T)$, then, reaches *eight* times the height of $\rho(f, T)$. It also spreads over twice the lateral extent, so the area under the curve, corresponding to the total energy radiated, increases *sixteenfold* on doubling the temperature: Stefan's Law, $P = \sigma T^4$.

UNDERSTANDING THE BLACK BODY CURVE

These beautifully precise experimental results were the key to a revolution. The first successful theoretical analysis of the data was by Max Planck in 1900. He concentrated on modeling the oscillating charges that must exist in the oven walls, radiating heat inwards and—in thermodynamic equilibrium—themselves being driven by the radiation field. The bottom line is that he found he could account for the observed curve *if* he required these oscillators not to radiate energy continuously, as the classical theory would demand, but *they could only lose or gain energy in chunks*, called *quanta*, of size hf, for an oscillator of frequency f. The constant h is now called Planck's constant, $h = 6.626 \times 10^{-34}$ joule.sec.

With that assumption, Planck calculated the following formula for the radiation energy density inside the oven:

$$\rho(f, T)df = \frac{8\pi V f^2 df}{c^3} \frac{hf}{e^{hf/kT} - 1}.$$

The perfect agreement of this formula with precise experiments, and the consequent necessity of energy quantization, was the most important advance in physics in the century. But no-one noticed for several years! His black body curve was completely accepted as the correct one: more and more accurate experiments confirmed it time and again, yet the radical nature of the quantum assumption didn't sink in. Planck wasn't too upset—he didn't believe it either, he saw it as a technical fix that (he hoped) would eventually prove unnecessary. Part of the problem was that Planck's route to the formula was long, difficult and implausible—he even made contradictory assumptions at different stages, as Einstein pointed out later. But the result was correct anyway, and to understand why we'll follow another, easier, route initiated (but not successfully completed) by Lord Rayleigh in England.

RAYLEIGH'S SOUND IDEA: COUNTING STANDING WAVES

In 1900, actually some months before Planck's breakthrough work, Lord Rayleigh was taking a more direct approach to the radiation inside the oven: he didn't even think about oscillators in the walls, he just took the radiation to be a collection of standing waves in a cubical enclosure: electromagnetic oscillators. In contrast to the somewhat murky reality of the wall oscillators, these standing electromagnetic waves were crystal clear.

This was a natural approach for Rayleigh he'd solved an almost identical problem a quarter century earlier, an analysis of standing *sound* waves in a cubical room (§267 of his book). The task is to find and enumerate the different possible standing waves in the room/oven, compatible with the boundary conditions. For sound waves in a room, the amplitude of the sound goes to zero at the walls. For the electromagnetic waves, the electric field parallel to the wall must go to zero if the wall is a perfect conductor (and it's OK to assume this see note later). So what are the allowed standing waves? The different allowed modes of vibration, that is, standing waves, in a string of length *a* fixed at both ends:

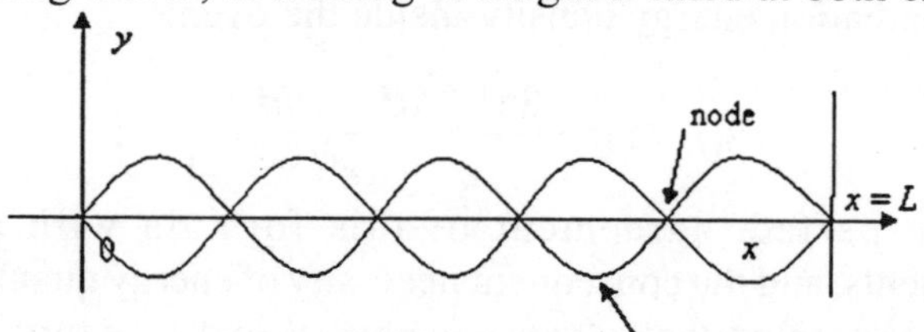

Fig. Possible mode of vibration of string with both ends fixed: $\lambda = 2L/5$

The possible values of wavelength are:

$$\lambda = 2a,\ a,\ 2a/3, \ldots$$

So the allowed frequencies are

$$f = c/\lambda = c/2a,\ 2(c/2a),\ 3(c/2a), \ldots$$

These allowed frequencies are equally spaced $c/2a$ apart. We define the spectral density by stating that number of modes between f and $f + \Delta f = N(f)\Delta f$ where we assume that Δf is large compared with the spacing between successive frequencies. Evidently for this one-dimensional exercise $N(f)$ is a constant equal to $2a/c$, each mode corresponds to an integer point on the real axis in units $c/2a$.

The amplitude of oscillation as a function of time is:

$$y = A \sin \frac{2\pi x}{\lambda} \sin 2\pi f t$$

more conveniently written.

$y = A \sin kx \sin \omega t$, where $k = 2\pi/\lambda$, $w = 2\pi f$, so $\omega = ck$.

The allowed values of k (called the wave number) are;

$$k = 2\pi/\lambda = \pi/a, 2\pi/a, 3\pi/a, \ldots \text{ and } f = ck/2\pi.$$

The generalization to three dimensions is simple: in a cubical box of side a, an allowed standing wave must satisfy the boundary conditions in all three directions. This means the choices of wave numbers are:

$$k_x = 2\pi/\lambda_x = \pi/a,\ 2\pi/a,\ 3\pi/a, \ldots$$
$$k_y = 2\pi/\lambda_y = \pi/a,\ 2\pi/a,\ 3\pi/a,$$
$$k_z = 2\pi/\lambda_z = \pi/a,\ 2\pi/a,\ 3\pi/a, \ldots$$

That is to say, each modes is labeled with three positive integers:

$$(k_x, k_y, k_z) = \frac{\pi}{a}(l, m, n)$$

and the frequency of the mode is:

$$f = ck/2\pi = (c/2\pi)\sqrt{k_x^2 + k_y^2 + k_z^2}\ .$$

For infrared and visible radiation in a reasonable sized oven, frequency intervals measured experimentally are far greater than the spacing $c/2a$ of these integer points. Just as in the one-dimensional example, these modes fill the three-dimensional k-space uniformly, with density $(a/\pi)^3$, but now this means the mode density is *not* uniform as a function of frequency.

The number of them between f and $f + \Delta f = N(f)\Delta f$ is the volume in k-space, in units $(\pi/a)^3$, of the spherical shell of radius $k = 2\pi f/c$, thickness $\Delta k = 2\pi\Delta f/c$, and restricted to all components of k being positive (like the integers), a factor of 1/8. Including a factor of 2 for the two polarization states of the standing electromagnetic waves, the density of states as a function of frequency in an oven of volume $V = a^3$ is:

$$N(f)\Delta f = \frac{1}{8} \times 2 \times \frac{4\pi k^2 \Delta k}{(\pi/a)^3} = \frac{1}{4} \times \left(\frac{a}{\pi}\right)^3 \times 4\pi\left(\frac{2\pi}{c}\right)^3 f^2 \Delta f$$

giving the density of radiation states in the oven

$$N(f)\Delta f = \frac{8V\pi f^2 \Delta f}{c^3}.$$

(Details of this analysis can be found in the notes. If you're wondering why it's OK to have an oven with essentially perfectly reflecting walls when we were previously insisting on absorbing walls, Kirchhoff proved long before that two such ovens at the same temperature will have the same radiation intensity—otherwise energy could be transferred from one to the other, violating the Second Law.)

WHAT ABOUT EQUIPARTITION OF ENERGY?

A central result of classical statistical mechanics is the equipartition of energy: for a system in thermal equilibrium, each degree of freedom has average energy $\frac{1}{2}kT$. Thus molecules in a gas have average kinetic energy $3/2kT$, $\frac{1}{2}kT$ for each direction, and a simple one-dimensional harmonic oscillator has total energy kT: $\frac{1}{2}kT$ kinetic energy and $\frac{1}{2}kT$ potential energy.

Comparing now the formula for the number of modes $N(f)\Delta f$ in a small interval Δf

$$N(f)\Delta f = \frac{8V\pi f^2 \Delta f}{c^3}$$

with Planck's formula for radiation energy intensity in the same interval:

$$\rho(f, T)\Delta f = \frac{8V\pi f^2 \Delta f}{c^3}\frac{hf}{e^{hf/kT}-1}$$

For the low frequency modes $hf \ll kT$ we can make the approximation

$$e^{hf/kT} - 1 \cong hf/kT$$

and it follows immediately that *each mode has energy kT*.

But things go badly wrong at high frequencies! The number of modes increases without limit, the *energy* in these high frequency modes, though, is decaying exponentially as the frequency increases. Ehrenfest later dubbed this the ultraviolet catastrophe. Rayleigh's sound approach apparently wasn't so sound after all something crucial was missing.

It is perhaps surprising that *Planck* never mentioned equipartition. Of course, as Rayleigh himself remarked, equipartition was well-known to have problems, for example in the specific heat of gases. And in fact Planck wasn't even sure about the *existence* of atoms: he later wrote that in the 1890's "I had been inclined to reject atomism". In fact, even Boltzmann was very unsure how well oscillators came to thermal equilibrium with electromagnetic radiation—after all, it was well known that oscillation of diatomic molecules failed to reach classical thermal equilibrium with kinetic energy.

(As long ago as 1877, Maxwell had pointed out that hot gases emit light at particular frequencies.

The frequencies do not change with temperature, so the oscillations must be simple harmonic but such an oscillator would surely also be excited by collisions at *low* temperatures, so why was energy not being fed into this mode?)

EINSTEIN SEES A GAS OF PHOTONS

After Planck announced his result in December, 1900 there was a deafening silence on the subject for several years. No-one (including Planck) realized the importance of what he had done his work was widely seen as just a clever technical fix, even if it did give the right answer (the curve itself was completely accepted as correct). Then in March, 1905, Albert Einstein turned his attention to the problem. He first rederived the Rayleigh result assuming equipartition:

$$\rho(f, T) = \frac{8\pi f^2}{c^3} kT$$

and observed that this made no sense at high frequencies. So he focused on Planck's formula for high frequencies, $hf >> kT$:

$$\rho(f, T)df = \frac{8V\pi f^3 df}{c^3} e^{-hf/kT}$$

(actually identical in this region to an earlier formula by Wien).

Einstein perceived an analogy here with the energy distribution in a classical gas. The (normalized) probability distribution function for classical atoms as a function of speed v was

$$f(v) = 4\pi\left(\frac{m}{2\pi kT}\right)^{3/2} v^2 e^{-B/kT},$$

and the corresponding energy density in v is

$$f(E) = \left[4\pi\left(\frac{m}{2\pi kT}\right)^{3/2} v^2\right] Ee^{-B/kT}$$

The radiation formula at high frequencies is

$$\rho(f, T) = \frac{8\pi f^2 hf}{c^3} e^{-hf/kT} ..$$

Einstein pointed out that if the high frequency radiation is imagined to be a gas of independent particles having energy $E = hf$, the energy density in frequency in the radiation is

$$\rho(E, T) = \left[\frac{8\pi f^2}{c^3}\right] Ee^{-E/kT} .$$

Comparing this with the expression for atoms, the analogy is close: recall that for the radiation, frequency is proportional to wave number and, on quantization, to momentum; for the (nonrelativistic) atoms velocity is proportional to momentum, so both these distributions are essentially in momentum space. Of course, the normalization factors differ, because the total number of atoms doesn't change with temperature, unlike the total radiation. Nevertheless, the analogy is compelling, and led Einstein to state that the radiation in the enclosure was itself quantized, the energy quantization was not some special property only of the wall oscillators, as Planck thought.

The radiation quanta are of course photons, but that word wasn't coined until later.Einstein had been troubled by Planck's derivation of his result, depending as it did first, on a classical analysis of the interaction between the wall oscillator and radiation, followed by a claim that the interaction was in fact not like that at all. But the answer was right, and now Einstein began to see why. In contrast to the poorly understood wall oscillators, the electromagnetic standing wave oscillations in the oven were completely clear.

ENERGY IN AN OSCILLATOR AS A FUNCTION OF TEMPERATURE

Einstein realized that, in terms of Rayleigh's electromagnetic standing waves, the blackbody radiation curves have a simple interpretation: the average energy in an oscillator of frequency f at temperature T is.

$$\vec{E} = \frac{hf}{e^{hf/kT} - 1}$$

Furthermore, Planck's work made plausible that this same quantization held for the material oscillators in the walls.

Einstein took the next step: he conjectured that all oscillators are quantized, for example a vibrating atom in a solid. This would explain why the Dulong Petit law, which assigns specific heat $3k$ to each atom in a solid, does not hold good at low temperatures: once $kT << hf$, the modes are not excited, so absorb little heat. The specific heat falls, as is indeed observed. Furthermore, it explains why diatomic gas molecules, such as oxygen and nitrogen, do not appear to absorb heat into vibrational modes these modes have very high frequency.

It's worth thinking about the constant exchange of energy with the environment for an oscillator in thermal equilibrium at temperature T. The random thermal fluctuations in a system have energy of order kT, this is the amount of energy, approximately, delivered back and forth.

But if an oscillator has $hf = 5kT$, say, it can only accept chunks of energy of size $5kT$, and will only be excited in the unlikely event that five of these random kT fluctuations come together at the right place at the right time. The high frequency modes are effectively frozen out by this minimum energy requirement. The exponential drop off in excitation with frequency reflects the exponential drop off in probability of getting the right number of fluctuations together, analogous to the exponential drop off in probability of tossing a coin n heads in a row.

SIMPLE DERIVATION OF PLANCK'S FORMULA FROM THE BOLTZMANN'S DISTRIBUTION

Planck's essential assumption in deriving his formula was that the oscillators only exchange energy with the radiation in quanta hf. Einstein made clear that the well-understood standing electromagnetic waves, the radiation in the oven, also have quantized energies.

The probability of a system at temperature T having energy E is proportional to $e^{-E/kT}$, Boltzmann's formula. It turns out that this formula continues to be valid in quantum systems. Now, a classical simple harmonic oscillator at T will have a probability distribution proportional to $e^{-E/kT} = e^{-(mv^2 + m\omega^2 x^2)/2kT}$, so the expectation value of the energy is

$$\vec{E} = \frac{\iint\left(\frac{1}{2}mv^2 + \frac{1}{2}m\omega^2x^2\right)e^{-(mv^2+m\omega^2x^2)/2kT}\,dvdx}{\iint e^{-(mv^2+m\omega^2x^2)/2kT}\,dvdx} = kT,$$

just the classical equipartition of energy.

But we now know this isn't true if the oscillator is quantized: the energies are now in steps hf apart.

Taking the ground state as the zero of energy, allowed energies are:

$$0, hf, 2hf, 3hf, \ldots$$

and assuming the Boltzmann expression for relative probabilities is still correct, the relative probabilities of these states will be in the ratios:

$$1: e^{-hf/kT}: e^{-2hf/kT}: e^{-3hf/kT}:\ldots$$

To find the oscillator energy at this temperature, we use these probabilities weighted by the corresponding energy, and divide by a normalization factor to ensure that the probabilities add up to 1:

$$\vec{E} = \frac{hfe^{-hf/kT} + 2hfe^{-2hf/kT} + 3hfe^{-3hf/kT}\ldots}{1 + e^{-hf/kT} + e^{-2hf/kT} + e^{-3hf/kT}\ldots}$$

$$= \frac{hf}{e^{hf/kT} - 1}$$

This is indeed the correct result from the black body experiments. Evidently Boltzmann's relative probability function $f^3/(e^{hf/kT} - 1)$ is still valid in quantum systems.

A NOTE ON WIEN'S DISPLACEMENT LAW

It is easy to see how Wien's Displacement Law follows from Planck's formula: the maximum radiation per unit frequency range is at the frequency f for which the function is a maximum. Solving numerically gives hf_{max}=2.82kT.

It can be established theoretically (and is confirmed experimentally) that the equation connecting the frequency of maximum energy intensity in units of Joules/m^3/Hz is:

$$f_{max} = 5.88 \times 10^{10} T \text{ Hz/K}.$$

However, the law is often stated in terms of the *wavelength* at which the intensity, now measured in Joules/m^3/m, that is, per unit interval of wavelength, and

$$\lambda_{max} = \frac{2.9 \times 10^{-3}}{T} m.K.$$

The important point to notice here is that these formulas do not give the same result, as is easily verified, since $\lambda_{max} \cong 1.7 \times$ 108 m/sec, not the speed of light!

The reason is that the two measures, per unit interval of frequency and per unit interval of wavelength, are different, so a claim that, say, sunlight is most intense in the yellow has to specify which is being used (actually it would be wavelength, frequency would give the near infrared).

BLACKBODY THERMODYNAMIC DERIVATION OF STEFAN'S LAW

In 1884, Poynting found expressions for the energy density, energy flow and momentum density and flow in an electromagnetic field: the energy density is $u = \frac{1}{2}\left(\varepsilon_0 \vec{E}^2 + \vec{B}^2 / \mu_0\right)$ the energy flow rate is given by the Poynting vector, $\vec{S} = (\vec{E} + \vec{B}) / \mu_0$, and the momentum density $\vec{P} = \varepsilon_0 (\vec{E} + \vec{B})$. (We write these in MKS, he used cgs.)

For a plane wave, $B = E/c$, ($c = 1/\sqrt{\varepsilon_0 \mu_0}$) so the energy density $u = \varepsilon_0 \vec{E}^2$ and the momentum density $|\vec{P}| = \varepsilon_0 \vec{E}^2 / c$. This was all deduced in the 1880's from a fairly lengthy (but straightforward, see for example Jackson's book) analysis of energy and momentum interchange between the electromagnetic field and charged particles.

It's easier to understand, though, using Einstein's (later) discovery of $E = mc^2$: the energy density u corresponds to a mass density $\rho = u/c^2$, the momentum density for a plane wave is then $|\vec{P}| = \rho c = u / c$. If we think of a light wave as having a finite total length, it follows that the energy-momentum relation must be

$$E = cp.$$

(Of course, this follows immediately from special relativity for massless particles, the point here is that it was previously known in the 1880's from electromagnetic theory.)

Now imagine an oven filled with radiation, energy density u, and consider the pressure on a wall. The argument is the same as for a gas of particles, except that the radiation travels at c and has energy-momentum relation $E = cp$. Taking the radiation to be perfectly reflected, and all directions to be equally likely, one finds

$$P = \frac{1}{3}u.$$

(Imagine first the radiation to be all perpendicular to a wall: in the steady state, there would be momentum density $u/2c$ in each direction (towards the wall and away from it). A quantity $uc/2c$ would bounce off the wall each second, for a total momentum change (hence pressure!) of u. In fact, only the perpendicular component of the radiation (which is going equally in all directions) counts, this gives the 1/3, exactly as for particles in the kinetic theory of gases.)

Another point: can we say the oven walls are perfectly reflecting?

Didn't we say they were black? Pippard points out that two such different ovens must, at the same temperature, have the same radiation distribution inside, or one could put a small pipe from one to the other and build a temperature difference, contradicting the Second Law.

At this point, we're ready to apply standard thermodynamics to the radiation. Assume the oven filled with radiation is equivalent to a classical gas in a piston (so the volume can be adjusted). Following Pippard (Classical Thermodynamics), the fundamental equation

$$dU = TdS - PdV$$

can be written in terms of energy and entropy *densities* $u = U/V$, $s = S/V$, and remembering $P = \frac{1}{3}u$, as

$$du = Tds + \frac{1}{V}\left(Ts - \frac{4}{3}u\right)dV.$$

But u and s only depend on temperature, not volume, so

$$du = Tds$$

and therefore

$$Ts = \frac{4}{3}u\,.$$

Hence

$$\frac{du}{ds} = T = \frac{\frac{4}{3}u}{s}$$

so $$u = \text{constant} \times S^{4/3}$$

from which

$$u = aT^4,$$

completing the thermodynamic derivation of Stefan's law.

THERMODYNAMIC DERIVATION OF WIEN'S DISPLACEMENT LAW

The idea here is to expand a cubical container isotropically and slowly (so thermal equilibrium is maintained), allowing no heat transfer (adiabatic expansion).

Imagining the radiation inside as standing waves in a reflecting box, the wavelengths will scale upwards with the dimensions of the box, so if we track waves of a particular wavelength, the wavelength will change as

$$\lambda^3 \propto V.$$

(this argument is from Wannier's book).

Now

$$PV = \frac{1}{3}U$$

and in the adiabatic expansion $dS = 0$, so

$$dU = -PdV$$

from which it is easy to check that the equation of the adiabat is

$$P^{3/4}V = \text{constant}$$

and since the radiation pressure goes as T^4,

$$T^3V = \text{constant.}$$

Suppose now we track waves having frequency in the small interval $(f_1, f_1 + df_1)$ as the volume increases slowly from V_1 to V_2.

The temperature will go from T_1 to T_2, the frequency range from (f_1, f_1+df_1) to (f_2, f_2+df_2) and from $\lambda f = c$,

$$f_1 = \frac{T_1}{T_2} f_2, \quad df_1 = \frac{T_1}{T_2} df_2.$$

Now we know from the Stefan-Boltzmann law that the total radiation intensity changes as:

$$\int_0^\infty \rho(f_1, T_1) df_1 = \frac{T_1^4}{T_2^4} \int_0^\infty \rho(f_2, T_2) df_2$$

Wien argued that Boltzmann's thermodynamic analysis could be applied to just the radiation in a small frequency interval, so one can conclude that

$$\rho(f_1, T_1) df_1 = \frac{T_1^4}{T_2^4} \rho(f_2, T_2) df_2$$

Changing notation: T_1 goes to T, T_2 goes to 1, f_1 goes to f, we have

$$\rho(f, T) = T^3 \rho\left(\frac{f}{T}, 1\right)$$

so apart from the overall factor T^3, the curve has exactly the same shape as a function of f/T at all temperatures. This is exactly what is observed experimentally. In particular, if the radiation is most intense at frequency f_{max}, then we find Wien's Displacement Law:

$$f_{max} \propto T.$$

The curve can equivalently be written

$$\rho(f, T) = f^3 g\left(\frac{f}{T}, 1\right),$$

probably this form inspired Wien's Radiation Law conjecture, $\rho(f) = \alpha f^3 e^{-\beta f/T}$.

THE FACTOR ¼ FOR RADIATION FROM A HOLE

If one imagines an energy density $\rho(f, T)$ streaming down a pipe of cross-section area A (the size of the hole) at speed c, one would find radiation intensity $Ac\,\rho(f, T)$. So why the factor of ¼? The energy density inside the oven has waves equally moving in opposite directions, so only half of it is moving the right way to get through the hole. Furthermore, it's not all coming directly at the hole, it's hitting it from all inside directions, and that coming in at an angle θ to the vertical sees effectively a smaller hole, by a factor $\cos\theta$. Taking account of this, and adding radiation from all angles, gives the ¼.

HOW PLANCK RELATED THE ENERGY DENSITY IN THE RADIATION FIELD TO THAT IN A SINGLE OSCILLATOR

He chose the simplest model possible: one linear oscillator (in the wall of the oven) driven at frequency $f = 2\pi\omega$ by a monochromatic oscillating electric field, the oscillator constantly damped by emitting energy as radiation.

It was well known from classical electrodynamics that an accelerating charge lost energy in this way, and it could be included in the oscillator dynamics as an effective damping force $-(2e^2/3c^3)\dddot{x}$, so the equation of motion is

$$m\ddot{x} + m\omega_0^2 x - (2e^2/3c^3)\dddot{x} = eE\cos\omega t$$

$m\omega_0^2$ being the usual spring constant. (*Note*: these are Planck's cgs units. In MKS, we should replace e^2 by $e^2/4\pi\varepsilon_0$, but we won't bother to add that clutter, since the e^2 terms cancel out in the final result.)

This can be further simplified by the following trick: the oscillator is driven at the external frequency $f = 2\pi\omega$, one can (at least for small damping) write

$$-(2e^2/3c^3)\dddot{x} = -(2e^2/3c^3)\omega^2\dot{x} = -\gamma\dot{x},$$

say. This is now a familiar equation, an oscillator in a viscous medium with a drag force proportional to velocity, and is readily solved, most simply by using complex numbers. The amplitude A of the driven oscillations is:

$$A = \frac{eE}{\sqrt{m^2(\omega_0^2 - \omega^2)^2 + (\gamma\omega)^2}},$$

The oscillator has energy $U = \frac{1}{2}m^2\omega_0^2 A^2$. For small damping, there is a sharp maximum at the natural frequency of the oscillator, ω_0.

To summarize: the physics of the above model is that the charged oscillator is being driven by the single-frequency external electromagnetic field, which is therefore feeding energy into the oscillator, which energy is constantly being drained out by radiation damping, that is, emission of outgoing radiation at the same frequency so in the steady state the oscillator is in equilibrium with the external field.

Planck's next step was to place this oscillator in a real radiation field, so the single frequency driving source is replaced by a very large number of driving fields at all different frequencies, and with no overall coherence. (If we take the oven to have reflective walls, there will be one driving field for each allowed standing wave, counting also a factor of two for polarization.) But there's a big simplification: if we take the damping to be small, the only *significant* driving frequencies are those close to the natural frequency of the oscillator, where

$$\begin{aligned} U &= \frac{1}{2} m^2 \omega_0^2 A^2 \\ &= \frac{1}{2} m\omega_0^2 \frac{e^2 E^2}{m^2(\omega_0^2 - \omega^2)^2 + (\gamma\omega)^2} \\ &\cong \frac{1}{2} m\omega_0^2 \frac{e^2}{4m^2\omega_0^2(\omega_0^2 - \omega^2)^2 + (\gamma\omega_0)^2} E^2 \\ &= \frac{1}{8m} \frac{e^2}{(\omega_0^2 - \omega^2)^2 + (\gamma/2m)^2} E^2. \end{aligned}$$

Now, going from a single driving field to an actual radiation field amounts to summing over the modes inside the narrow peak at ω_0. The weight can be found by integration: it's

$$\frac{\pi e^2}{8m}\frac{2m}{\gamma} = \frac{\pi e^2}{4}\frac{3c^3}{2e^2\omega_0^2} = \frac{3\pi c^3}{8\omega_0^2}$$

from which:

$$E^2 = \frac{8\omega_0^2}{3\pi c^3} U$$

Now this is only the x-component of the radiation field, going to the full field gets rid of the 3 in the denominator. The energy density in (radial) frequency in the radiation field is then found by replacing

$$E^2/2 \rightarrow 4\pi\rho(\omega, T)$$

giving

$$\rho(\omega_0, T) = \frac{\omega_0^2}{3\pi^2 c^3} U,$$

or in terms of frequency in cycles per second,

$$\rho(f, T) = (8\pi f^2/c^3)\, U(f, T)$$

where $U(f, T)$ is the energy of the oscillator, $\rho(f, T)$ the radiation energy intensity.

RADIATION DAMPING AND REVERSIBILITY

It is perhaps worth noting that although the radiation term is *formally* equivalent to a viscous damping term, it does not produce "heat" in quite the same way the energy lost goes back into the driving system, and the damping constant is not arbitrary, but related to the driving coupling.

Planck believed at first that this absorption of radiation and reemission was irreversible, in line with the Second Law. But Boltzmann pointed out to him that the electromagnetic processes were reversible, just as mechanical processes were, so this couldn't be quite right.

COUNTING THE ELECTROMAGNETIC STANDING WAVE MODES IN THE OVEN

To make any *quantitative* progress in analyzing the radiation, we must have a clear picture of the degrees of freedom of this system: how many oscillators have frequencies in a given energy range?

These oscillations are standing electromagnetic waves. The waves are contained in the oven, so the electric field intensity drops rapidly to zero on approaching and going into the walls, because inside the walls the electric energy will be rapidly dissipated by currents or polarization.

In fact, then, the boundary condition at the walls is much like that for waves on a string fixed at both ends, where the wave amplitude goes to zero at the ends.

The frequency distribution function of the possible different modes of vibration (that is, the different degrees of freedom) of a string stretched between two points a distance L apart.

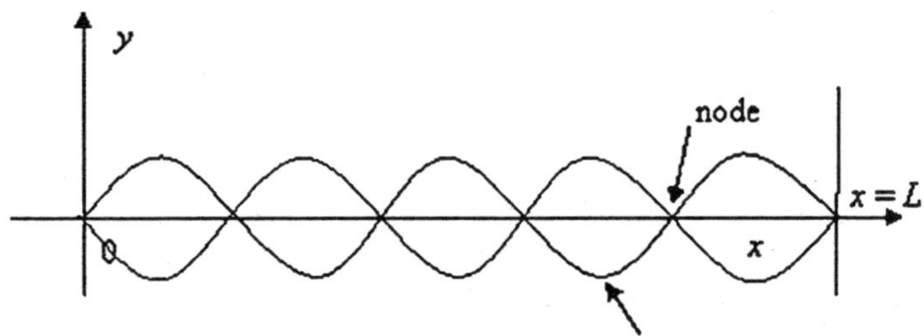

Fig. Possible mode of vibration of string with both ends fixed: $\lambda = 2L/5$

Possible values of the wavelength are:

$$\lambda = 2L, L, 2L/3, \ldots$$

so the frequency

$$f = c/\lambda = c/2L, 2(c/2L), 3(c/2L), \ldots$$

The allowed frequencies are thus equally spaced $c/2L$ apart. We define the spectral density by stating that the number of modes between f and $f + \Delta f$ is $N(f)\Delta f$ for Δ small compared with the range of frequencies in the system, but large compared with the spacing between successive allowed frequencies.

Evidently, in this simple one-dimensional example $N(f)$ is a constant equal to $2L/c$.

The amplitude of oscillation as a function of time has the form:

$$y = A \sin \frac{2\pi x}{\lambda} \sin 2\pi f t.$$

This is more conveniently written:

$$y = A \sin kx \sin \omega t$$

where

$$k = 2\pi/\lambda,\ \omega = 2\pi f = ck$$

using $\lambda f = c$.

Here w is of course the *radial* frequency—how many *radians* per second, rather than cycles per second, the time dependent component of the wave moves through. Analogously, k, called the *wave number*, counts the radians of spatial change in unit length (one meter in SI, but note that practicing physicists often still use CGS).

We are now ready to tackle the more complex problem of three-dimensional standing electromagnetic waves in a cubical oven. The first guess, just generalizing the expression above, would be:

$$\vec{E}(\vec{r},t) = \vec{E}_0 \sin k_x x \sin k_y \sin k_z \sin \omega t.$$

This will satisfy the wave equation

$$\vec{\nabla}^2 \vec{E} - \frac{1}{c^2}\frac{\partial^2 \vec{E}}{\partial t^2} = 0$$

if

$$\omega^2 = c^2 (k_x^2 + k_y^2 + k_z^2).$$

Also the electric field will be zero at all the walls if we choose the k's appropriately, that is, so that $\sin k_x L = 0$, etc.

However, this wave is not Quite Correct!

There is no electric charge in the oven away from the walls, so the divergence of the electric field, $\vec{\nabla} \cdot \vec{E} = \rho/\varepsilon_0$, must be identically zero away from the walls, a condition this wave *does not satisfy*. The divergence is:

$$\begin{aligned}\vec{\nabla} \cdot \vec{E}(\vec{r},t) = {} & E_{0x} k_x \cos k_x x \sin k_y y \sin k_z z \sin \omega t \\ & + E_{0z} k_y \cos k_x x \sin k_y y \sin k_z z \sin \omega t \\ & + E_{0z} k_z \cos k_x x \sin k_y y \sin k_z z \sin \omega t\end{aligned}$$

and, for example, taking any x for which $\sin k_x x = 0$, the second and third terms above are identically zero, but the first term certainly isn't.

This problem can be fixed by thinking more carefully about the boundary conditions at the walls. It is true that any component of the electric field *parallel* to the wall will be attenuated rapidly by currents or polarization in the wall. However, there could be an electric field *perpendicular* to the wall, because there could be *surface charge* on the wall.

This means a possible solution is:

$$E_x(\vec{r},t) = E_{0x} k_x \cos k_x x \sin k_y y \sin k_z z \sin \omega t$$

$$E_y(\vec{r},t) = E_{0y} \sin k_x x \cos k_y y \sin k_z z \sin \omega t$$
$$E_z(\vec{r},t) = E_{0z} \cos k_x x \sin k_y y \sin k_z z \sin \omega t$$

These components are deliberately chosen to give

$$\vec{\nabla} \cdot \vec{E}(\vec{r},t) = (k_x E_{0x} + k_y E_{0y} + k_z E_{0z}) \sin k_x x \sin k_y y \sin k_z z \sin \omega t$$

so choosing $\vec{k}$ such that $\vec{k} \cdot \vec{E}_0 = 0$, we find $\vec{\nabla} \cdot \vec{E} = 0$

This is telling us something we should have realized from the beginning: in the standing electromagnetic wave in the cavity, just as in a propagating wave, the electric field is perpendicular to the direction in which the wave is moving. In other words, it is a transverse wave, and in fact there are two independent polarizations, which we must remember to count when we find the total number of degrees of freedom.

It remains to find the allowed frequencies of vibration—the normal modes—of the electromagnetic radiation in the oven. The first step is to find the allowed values of the wave numbers k_x, k_y and k_z. These are fixed by the boundary conditions $\sin k_x L = 0$, etc., so

$$k_{x,y,z} = \pi/L, 2\pi/L, 3\pi/L,$$

so the vector

$$\vec{k} = (k_x, k_y, k_z) = \frac{\pi}{L}(l, m, n)$$

where(l, m, n) are positive integers, like (1, 1, 1) or (5, 13, 4): in other words, each possible standing electromagnetic wave in the oven corresponds to a point in the (k_x, k_y, k_z) space labeled by three positive integers. These are the *intersection points of a cubic lattice* in the space $k_x > 0$, $k_y > 0$, $k_z > 0$.

Each such point can be associated with the little lattice cube of volume $(\pi/L)^3$ for which the point in question is the furthest corner of the little cube from the *k*-space origin. These little cubes stack together to fill all that part of *k*-space having $k_x > 0$, $k_y > 0$ and $k_z > 0$.

The frequency of vibration of the wave having wave numbers (k_x, k_y, k_z) is

$$\omega = c\sqrt{k_x^2, k_y^2, k_z^2}$$

Recall that the experimental measurement of black body radiation from the oven detects the intensity of radiation in a given *frequency* range. It does *not* tell us the wave numbers of the radiating modes!

Hence, to compare with these experimental results, we must choose a small frequency range and find how many possible sets of wave numbers correspond to modes of vibration having frequencies within that range. Notice that a fixed value of the frequency corresponds to a spherical surface in *k*-space:

$$\frac{\omega^2}{c^2} = k_x^2, k_y^2, k_z^2 .$$

Therefore, the number of possible modes of radiation in the oven having frequencies in the range $(\omega, \omega + \Delta\omega)$ is equal to the number of lattice-point wave number values (k_x, k_y, k_z) between two spherical surfaces centered at the origin and having radii $(\omega/c, (\omega + \Delta\omega)/c)$, and only counting in the positive octant, $k_x > 0$, $k_y > 0$ and $k_z > 0$.

Now, as we argues above, each lattice point can be associated with a small cube of volume $(\pi/L)^3$. Assuming we choose $\Delta\omega/c >> \pi/L$, there will be many of these small cubes between the spherical surfaces, and the total number of lattice points between the spheres in the positive octant will be just the volume of the space between the spheres divided by the volume of one of these cubes.

So the number of possible modes of radiation in the oven having frequencies in the range $(\omega, \omega + \Delta\omega)$ is:

$$\frac{(1/8) \times 4\pi(\omega/c)^2 \Delta(\omega/c)}{(\pi/L)^3} = (L^3/2\pi^2)(\omega/c)^2 \Delta(\omega/c)$$

Putting this in terms of the frequency f in cycles per second, $\omega = 2\pi f$, and *inserting an extra factor of two* for the two independent polarizations of the wave (discussed above) we find the distribution function for modes in frequency is given by:

$$N(f)df = \frac{8V\pi f^2 df}{c^3}$$

where $V = L^3$ is the volume of the oven.

PLANCK'S ROUTE TO THE BLACK BODY RADIATION FORMULA AND QUANTIZATION

WIEN'S RADIATION LAW

Wien proved using classical thermodynamics that the shape of the black body curve didn't change with temperature, the curve just grew and expanded. However, the thermodynamic methods didn't specify the actual shape. In 1893, Wien made a guess, based on the experimental data:

$$\rho(f) = \alpha f^3 e^{-\beta f/T},$$

his Radiation Law. α, β are constants. In fact, this provided an excellent fit: it seemed that Kirchhoff's challenge had at last been met! But the long wavelength low frequency measurements were not very precise, and when improved infrared technology was introduced a few years later, in 1900, it became clear that at the lowest frequencies $\rho(f)$ went as f^2, not f^3, and furthermore the radiation intensity at these low frequencies was proportional to the temperature. So Wien's formula was wrong. The challenge was still there.

PLANCK'S THERMODYNAMIC APPROACH: OSCILLATORS IN THE OVEN WALL

After what's happened in physics over the last century or so, it's difficult to appreciate the mindset of a physicist like Planck in the late 1890's. He was forty years old and a well-established theorist at the University of Berlin. His earlier work had been in chemical physics, where applying thermodynamics had led to brilliant successes.

He was convinced thermodynamics was the key to understanding nature at the deepest level. He spent years clarifying the subtleties of the Second Law (that entropy always increases). He believed the Second Law was rigorously correct, and would eventually be proved so in a more fundamental theory.

And now thermodynamics had made a good start in analyzing black body radiation, with proofs of Stefan's Law and Wien's Displacement Law.

It seemed very likely that thermodynamics would yield the whole black body radiation curve. He felt this curve was the key to understanding just how electromagnetic radiation and matter exchanged energy. This was one of the basic problems in physics, and of obvious technological importance. And, in fact, just at that time experimentalists at his university were measuring the black body radiation curve to new levels of precision.

(Boltzmann himself had gone on to a molecular analysis of the properties of gases, relating thermodynamic quantities to microscopic distributions of particles.

Planck was not impressed by this approach, since it implied that the Second Law was only statistical, only valid in the limit of large systems. However, he *was* fairly familiar with Boltzmann's work—he'd taught it in some of his classes, to present all points of view. Planck wasn't sure, though, that atoms and molecules even existed.)

But how to begin a thermodynamic analysis of black body radiation? The oven used by the experimentalists was a dauntingly complex system: the hot oven walls contained many tiny oscillating electrical charges, the electromagnetic radiation from the acceleration of these charges being the heat and light radiation in the oven.

At the same time, the wall oscillators were *supplied* with energy by the oscillating electrical fields of the radiation. In other words, at a steady temperature, the radiation inside the oven and the electrical oscillators in the walls were in *thermal equilibrium*, there was no net transfer of energy from one to the other over time, but small amounts of energy were constantly being traded back and forth for individual oscillators.

Fortunately, Kirchhoff had long ago proved that the details of the oven don't matter, if two ovens at the same temperature have different radiation intensity at some particular frequency, energy could flow from one to the other, violating the Second Law. So Planck could consider as his "oven" the simplest possible material object that would interact with the radiation: he took a simple harmonic oscillator (one-dimensional, mass m, linear restoring force $-m\omega_0^2 x$).

He replaced the incoherent heat radiation with a monochromatic oscillating electric field driving the oscillator. The model oscillator carried a charge e. It was well-established in electromagnetic theory that an accelerating charge loses energy by emitting radiation, the effective drag force being $(2e^2/3c^3)\dddot{x}$.

The oscillator equation of motion is therefore:

$$m\ddot{x} + m\omega_0^2 x - (2e^2/3c^3)\dddot{x} = eE\cos\omega t.$$

For the driven oscillator, Planck took $\dddot{x} = -\omega^2\dot{x}$, giving a standard classical mechanics problem: the driven simple harmonic oscillator with viscous damping. This he solved to find the energy in the oscillator in terms of the strength of the driving field.

A CLASSICAL RESULT RELATING RADIATION INTENSITY TO OSCILLATOR ENERGY

Planck next replaced the single driving field by the incoherent field of radiation in an oven in equilibrium at temperature *T*. This is a completely different scenario! With the single driving field, after initial transient behaviour, the oscillator settles down at a fixed amplitude and phase entrained to the driving force. But with many incoherent driving fields, even though the important ones turn out to be those close to the natural frequency of the oscillator, its motion will no longer much resemble a clock pendulum, more like an outdoor hanging basket in chaotic weather.

However, Planck was considering *energy exchange*, he was not interested in a detailed description of the motion; the energy in the oscillator goes as the square of the driving field, and with many incoherent fields driving, the total oscillator energy is just the sum from each separately (cross terms will average to zero).

He was able to establish from his analysis an important correspondence between the wall oscillator's *mean energy* $U(f, T)$ and the energy $\rho(f, T)$ density per unit frequency in the radiation field:

$$\rho(f, T) = (8\pi f^2/c^3)\, U(f, T).$$

It's worth emphasizing that this is a classical result: the *only* inputs are classical dynamics, and Maxwell's electromagnetic theory. Notice that the charge e of the oscillator doesn't appear: the result is independent of the coupling strength between the oscillator and the radiation, the coupling only has to be strong enough to ensure thermal equilibrium.

OSCILLATOR THERMODYNAMICS: PLANCK FOCUSES ON ENTROPY

Armed with this new connection between the black body curve and the energy of an oscillator, Planck realized that from Wien's Radiation Law,

$$\rho(f) = \alpha f^3 e^{-\beta f/T},$$

he could work out completely the thermodynamics of an oscillator. From his Second Law perspective, the natural approach was to see how the entropy varied with energy, the energy being given (from the two equations above) by

$$U(f) = \frac{\alpha c^3 f}{8\pi} e^{-\beta f/T}$$

He wrote down the corresponding expression for the entropy:

$$S = -\frac{U}{\beta f}\left(\ln\frac{8\pi U}{\alpha f c^3} - 1\right)$$

Exercise: Prove $TdS = dU$.

An important quantity in thermodynamics is the second derivative of the entropy, this is closely related to the Second Law: negativity of the second derivative guarantees that the entropy will increase back to equilibrium if the system is disturbed.

He found an elegant result:

$$\frac{\partial^2 S}{\partial U^2} = -\frac{1}{\beta f U}$$

guaranteed to be negative! He was impressed by this simplicity, and thought he must be close to a deep thermodynamic truth. He went on to argue that this indicated the Second Law plus the displacement law most likely determined the black body curve uniquely. He was wrong. Of course, it had been well known for years that statistical mechanics applied to an oscillator gives it energy kT. Why Planck didn't even *mention* this is a total mystery. He was familiar with Boltzmann's work, but he really didn't care for statistical mechanics. He was an old fashioned thermodynamics.

NEW EXPERIMENTS, NEW THEORY

In October 1900, Rubens and Kurlbaum in Berlin announced

some new experimental findings: the radiation intensity at low frequencies went as f^2, not f^3, and the low frequency intensity was proportional to the temperature.

This shook Planck. His simple result for an oscillator, $\partial^2 S/\partial U^2 = -1/\beta f U$ was still very close to the truth for high frequencies, but at low frequencies equipartition was holding, the oscillator energy being $U = kT$. Together with $dU = TdS$, this gave immediately $\partial^2 S/\partial U^2 = -k/U^2$.

This certainly meant that his previous argument that the curve was uniquely determined by $\partial^2 S/\partial U^2 = -1/\beta f U$ had to be wrong. Abandoning thoughts of deep thermodynamic truths, he decided he'd better patch things up as best he could. How do you get from $\partial^2 S/\partial U^2 = -k/U^2$ at low frequencies to $\partial^2 S/\partial U^2 = -1/\beta f U$ at high frequencies? Well, there's one simple way:

$$\partial^2 S/\partial U^2 = -\frac{k}{U(hf+U)}$$

(I've put in the correct values for the two parameters here: k is Boltzmann's constant, necessary to match the low frequency equation, the constant β in Wien's formula turns out to be Planck's constant divided by Boltzmann's constant, $\beta = h/k$.) This was of course a completely unjustified guess. But pressing on, integrating twice gives the entropy:

$$S = k[(1 + U/hf)\ln(1 + U/hf) - (U-hf)\ln(U/hf)]$$

from which

$$\frac{dS}{dU} = \frac{k}{hf}\left[\ln\left(1+\frac{U}{hf}\right) - \ln\frac{U}{hf}\right] = \frac{1}{T},$$

giving

$$U = \frac{hf}{e^{hf/kT}-1}.$$

This yields the radiation curve:

$$\rho(f, T) = \frac{8\pi f^2}{c^3}\frac{hf}{e^{hf/kT}-1}$$

(energy per unit volume) where k is Boltzmann's constant and h is *a new constant*, now known as Planck's constant.

This worked brilliantly! Of course, it matched Wien's formula for high frequencies, and was proportional to T at low frequencies. But it turned out to be *far better*: it matched the new high precision measurements within their tiny limits of error, throughout the entire range.

THE GREAT BREAKTHROUGH: BIRTH OF THE QUANTUM

This surprisingly good news had Planck desperately searching for some theoretical justification! As always, he focused on the entropy:

$$S = k[(1 + U/hf) \ln (1 + U/hf) - (U-hf) \ln(U/hf)]$$

How could this expression be interpreted? Here things took a very unexpected turn. Ironically, his familiarity with Boltzmann's analysis of the entropy of a gas of atoms provided the clue, even though Planck himself doubted the existence of atoms!

Boltzmann's expression for entropy S is

$$S = k \ln W$$

where W is the volume of microscopic phase space corresponding to given macroscopic variables, and k is Boltzmann's constant. (In fact this formulation, which appears on Boltzmann's grave, was first written down by Planck.)

In 1877, Boltzmann actually analyzed a model system of atoms having entropy very close to the expression Planck had found. Boltzmann's model allowed the atoms to have only energies which were integer multiples of a small fixed energy ε. He then found the number of possible arrangements of atoms corresponding to a given total energy. This combinatorial analysis, using Stirling's formula $\ln N! \cong N \ln N - N$, gave him expressions for the entropy. Finally, Boltzmann took the limit of small ε.

How would a similar analysis work for the harmonic oscillator? Again, one of Boltzmann's ideas proved useful. The entropy of an oscillator having mean energy U at temperature T is related to the volume of phase space the oscillator is knocked around in by the thermal noise interactions.

In 1884, Boltzmann had introduced the concept of an ensemble: for the oscillator, this would be a large collection of N identical oscillators, with random phases but all the same U and

T, so that at one instant in time the whole collection will represent the possible states of the single oscillator over time.

The ensemble then has entropy NS, where S is the entropy of the single oscillator. The entropy of the ensemble is $NS = Nk \ln W$. Now

$$S = kN[(1 + U/hf) \ln (1 + U/hf) - (U-hf) \ln(U/hf)],$$

so

$$\ln W = N[(1 + U/hf) \ln (1 + U/hf) - (U-hf) \ln(U/hf)]$$

The possible different arrangements of the N oscillators amount to: How many different ways can total energy NU be shared among the N oscillators?

That is what W measures. In fact, Planck realized, probably remembering Boltzmann's work, this expression for W closely resembles a well-known combinatorial expression: how many ways can M objects by distributed among N boxes, if we assume the objects are all identical? The answer is

$$W = \frac{(N + M - 1)!}{M!(N - 1)!}$$

Proof: Put M dots on a line, with N–1 vertical lines interspersed. These lines are the division from one box to the next. W above is the total number of ways of doing this, since the dots are all identical, and so are the lines.)

For large values of N, M, using Stirling's formula $N! \cong N \ln N - N$ this becomes

$$\ln W = N[(1 + M/N) \ln (1 + M/N) - (M-N) \ln(M/N)]$$

And, staring at this formula (or just remembering it?) then at his expression for W, it dawned on Planck that *they are the same* provided

$$U/hf = M/N, \text{ or } NU = Mhf.$$

That is to say, the total energy NU of Planck's array of N identical oscillators of frequency is Mhf, and—crucially—the entropy expression tells us this energy is distributed among the oscillators in discrete chunks each of size hf.

Planck did this work by December 1900, in two intense months after learning the new experimental results and feeling he had to justify his curve that fit so well. But he only half believed it. After all, the first part of his derivation, identifying the energy of an oscillator with that in the radiation field, was purely classical: he'd assumed the emission and absorption of energy to be continuous. Then, he suddenly changed the story, moving to a totally nonclassical concept, that the oscillators could only gain and lose energy in chunks, or quanta.

(Incidentally, it didn't occur to him that the radiation itself might be in quanta: he saw this quantization purely as a property of the wall oscillators.) Although the exactness of his curve was widely admired, and it *was* the Birth of the Quantum Theory (with hindsight), no-one—including Planck—grasped this for several years!

MEANWHILE IN ENGLAND

Lord Rayleigh was working on the same problem. He'd heard about Wien's Radiation Law $\rho(f) = \alpha f^3 e^{-\beta f/T}$ but he didn't believe it—for one thing, it predicted that if you detected radiation at one frequency as the oven heated up, beyond a certain temperature, according to Wien, the radiation intensity would not increase.

As Rayleigh pointed out in May, 1900, this seemed very implausible—and, in the infrared, the effect should be detectable by those excellent Berlin experimentalists who had (apparently!) confirmed Wien's Law. He was of course correct.

At the same time, he proposed changing the law from Wien's form to $\rho(f) = c_1 f^2 T e^{-c_2 f/T}$, Rayleigh's Radiation Law.

This would ensure that for a particular frequency, the radiation intensity at high enough temperature would become linear in T, a much more reasonable result, in accord with equipartition. He published a two page paper making this point in June, 1900. But his Law didn't last long: in October 1900, Rubens and Kurlbaum's very precise infrared measurements showed his predicted curve was just outside their error bars (the plot is ρ versus T in the infrared, wavelength λ = 51.2 microns. The solid line is observation, Rayleigh is dot-dash, note Wien is hopeless):

Rayleigh's rationale for the f^2 low frequency behaviour on Black Body Radiation: he took the radiation to be a collection of standing waves in a cubical enclosure: electromagnetic oscillators, and found all the allowed such standing waves for each frequency interval. He proved that the density of such modes of vibration as a function of frequency went as f^2. So assuming equipartition of energy, obviously nonsense at high frequencies, but as Rayleigh commented: "...although for some reason not yet explained the doctrine fails in general, it seems possible it may apply to the graver modes" meaning the low frequency modes. He was quite right. Rubens and Kurlbaum show Rayleigh's curve on their graph reproduced above. However, they give equal billing to some other curves, all of which are really guesses.

They didn't apparently realise that Rayleigh's curve was based on theory. More remarkably, Planck refers explicitly to their paper in his December 1900 work—yet does not mention Rayleigh's work. Perhaps this is a question of style—possibly Rayleigh's paper, just two pages long, called "Remarks", and with the appearance of having being written in an hour or two, didn't impress the Berliners. This was a pity, because Rayleigh's approach, based on standing electromagnetic waves, proved most fruitful when it was taken up, or perhaps rediscovered, by Einstein years later.

Chapter 7

Rays and Particles

THE PHOTOELECTRIC EFFECT

HERTZ FINDS MAXWELL'S WAVES: AND SOMETHING ELSE

The most dramatic prediction of Maxwell's theory of electromagnetism, published in 1865, was the existence of electromagnetic waves moving at the speed of light, and the conclusion that light itself was just such a wave. This challenged experimentalists to generate and detect electromagnetic radiation using some form of electrical apparatus.

The first clearly successful attempt was by Heinrich Hertz in 1886. He used a high voltage induction coil to cause a spark discharge between two pieces of brass, to quote him, "Imagine a cylindrical brass body, 3 cm in diameter and 26 cm long, interrupted midway along its length by a spark gap whose poles on either side are formed by spheres of 2 cm radius." The idea was that once a spark formed a conducting path between the two brass conductors, charge would rapidly oscillate back and forth, emitting electromagnetic radiation of a wavelength similar to the size of the conductors themselves.

To prove there really was radiation emitted, it had to be detected. Hertz used a piece of copper wire 1 mm thick bent into a circle of diameter 7.5 cms, with a small brass sphere on one end, and the other end of the wire was pointed, with the point near the sphere.

He added a screw mechanism so that the point could be moved very close to the sphere in a controlled fashion. This "receiver" was designed so that current oscillating back and forth in the wire would have a natural period close to that of the "transmitter" described above. The presence of oscillating charge in the receiver would be signaled by a spark across the (tiny) gap between the point and the sphere (typically, this gap was hundredths of a millimeter). (It was suggested to Hertz that this spark gap could be replaced as a detector by a suitably prepared frog's leg, but that apparently didn't work.)

The experiment was very successful - Hertz was able to detect the radiation up to fifty feet away, and in a series of ingenious experiments established that the radiation was reflected and refracted as expected, and that it was polarized. The main problem - the limiting factor in detectionwas being able to see the tiny spark in the receiver. In trying to improve the spark's visibility, he came upon something very mysterious.

To quote from Hertz again (he called the transmitter spark A, the receiver B): "I occasionally enclosed the spark B in a dark case so as to more easily make the observations; and in so doing I observed that the maximum spark-length became decidedly smaller in the case than it was before. On removing in succession the various parts of the case, it was seen that the only portion of it which exercised this prejudicial effect was that which screened the spark B from the spark A.

The partition on that side exhibited this effect, not only when it was in the immediate neighbourhood of the spark B, but also when it was interposed at greater distances from B between A and B. A phenomenon so remarkable called for closer investigation."

Hertz then embarked on a very thorough investigation. He found that the small receiver spark was more vigorous if it was exposed to ultraviolet light from the transmitter spark. It took a long time to figure this out - he first checked for some kind of electromagnetic effect, but found a sheet of glass effectively shielded the spark.He then found a slab of quartz did not shield the spark, whereupon he used a quartz prism to break up the light from the big spark into its components, and discovered that the wavelength which made the little spark more powerful was beyond the visible, in the ultraviolet.

In 1887, Hertz concluded what must have been months of investigation: " I confine myself at present to communicating the results obtained, without attempting any theory respecting the manner in which the observed phenomena are brought about."

HALLWACHS' SIMPLER APPROACH

The next year, 1888, another German physicist, Wilhelm Hallwachs, in Dresden, wrote:

"In a recent publication Hertz has described investigations on the dependence of the maximum length of an induction spark on the radiation received by it from another induction spark. He proved that the phenomenon observed is an action of the ultraviolet light. No further light on the nature of the phenomenon could be obtained, because of the complicated conditions of the research in which it appeared.

I have endeavored to obtain related phenomena which would occur under simpler conditions, in order to make the explanation of the phenomena easier. Success was obtained by investigating the action of the electric light on electrically charged bodies." He then describes his very simple experiment: a clean circular plate of zinc was mounted on an insulating stand and attached by a wire to a gold leaf electroscope, which was then charged negatively.

The electroscope lost its charge very slowly. However, if the zinc plate was exposed to ultraviolet light from an arc lamp, or from burning magnesium, charge leaked away quickly. If the plate was positively charged, there was no fast charge leakage. Questions for the reader: Could it be that the ultraviolet light somehow spoiled the insulating properties of the stand the zinc plate was on? Could it be that electric or magnetic effects from the large current in the arc lamp somehow caused the charge leakage?

Although Hallwach's experiment certainly clarified the situation, he did not offer any theory of what was going on.

J.J. THOMSON IDENTIFIES THE PARTICLES

In fact, the situation remained unclear until 1899, when Thomson established that the ultraviolet light caused *electrons* to be emitted, the same particles found in cathode rays. His method was to enclose the metallic surface to be exposed to radiation in a vacuum tube, in other words to make it the cathode in a cathode ray tube. The new feature was that electrons were to be ejected from the cathode by the radiation, rather than by the strong electric field used previously.

By this time, there was a plausible picture of what was going on. Atoms in the cathode contained electrons, which were shaken and caused to vibrate by the oscillating electric field of the incident radiation. Eventually some of them would be shaken loose, and would be ejected from the cathode.

It is worthwhile considering carefully how the *number* and *speed* of electrons emitted would be expected to vary with the *intensity* and *colour* of the incident radiation. Increasing the intensity of radiation would shake the electrons more violently, so one would expect more to be emitted, and they would shoot out at greater speed, on average.

Increasing the frequency of the radiation would shake the electrons faster, so might cause the electrons to come out faster. For very dim light, it would take some time for an electron to work up to a sufficient amplitude of vibration to shake loose.

LENARD FINDS SOME SURPRISES

In 1902, Lenard studied how the energy of the emitted photoelectrons varied with the intensity of the light. He used a carbon arc light, and could increase the intensity a thousand-fold. The ejected electrons hit another metal plate, the collector, which was connected to the cathode by a wire with a sensitive ammeter, to measure the current produced by the illumination.

To measure the energy of the ejected electrons, Lenard charged the collector plate negatively, to repel the electrons coming towards it.

Thus, only electrons ejected with enough kinetic energy to get up this potential hill would contribute to the current. Lenard discovered that there was a well defined minimum voltage that stopped any electrons getting through, we'll call it V_{stop}.

To his surprise, he found that V_{stop} did not depend at all on the intensity of the light! Doubling the light intensity doubled the *number* of electrons emitted, but did not affect the *energies* of the emitted electrons. The more powerful oscillating field ejected more electrons, but the maximum individual energy of the ejected electrons was the same as for the weaker field.

But Lenard did something else. With his very powerful arc lamp, there was sufficient intensity to separate out the colors and check the photoelectric effect using light of different colors. He found that the maximum energy of the ejected electrons *did* depend on the colorthe shorter wavelength, higher frequency light caused electrons to be ejected with more energy.

This was, however, a fairly qualitative conclusionthe energy measurements were not very reproducible, because they were extremely sensitive to the condition of the surface, in particular its state of partial oxidation.

In the best vacua available at that time, significant oxidation of a fresh surface took place in tens of minutes. The details of the surface are crucial because the fastest electrons emitted are those from right at the surface, and their binding to the solid depends strongly on the nature of the surfaceis it pure metal or a mixture of metal and oxygen atoms.

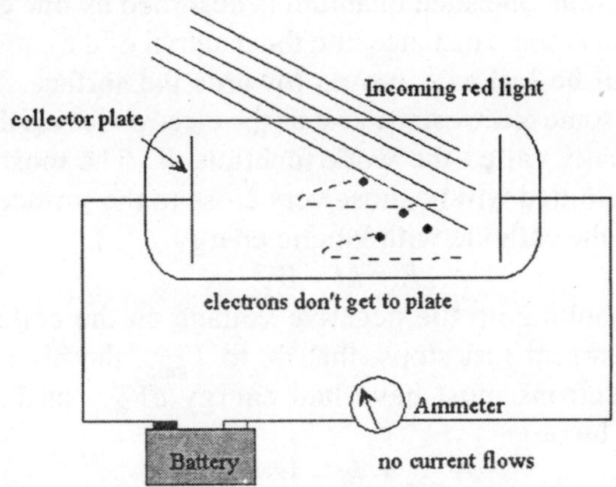

In the above figure, the battery represents the potential Lenard used to charge the collector plate negatively, which would actually be a variable voltage source.

Since the electrons ejected by the blue light are getting to the collector plate, evidently the potential supplied by the battery is less than V_{stop} for blue light. Show with an arrow on the wire the direction of the electric current in the wire.

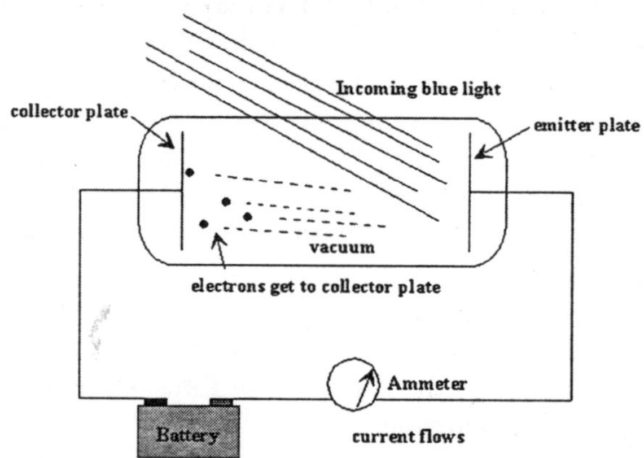

Einstein Suggests an Explanation

In 1905 Einstein gave a very simple interpretation of Lenard's results. He just assumed that the incoming radiation should be thought of as quanta of frequency *hf*, with *f* the frequency. In photoemission, one such quantum is absorbed by one electron. If the electron is some distance into the material of the cathode, some energy will be lost as it moves towards the surface. There will always be some electrostatic cost as the electron leaves the surface, this is usually called the work function, *W*. The most energetic electrons emitted will be those very close to the surface, and they will leave the cathode with kinetic energy

$$E = hf - W.$$

On cranking up the negative voltage on the collector plate until the current just stops, that is, to V_{stop}, the highest kinetic energy electrons must have had energy eV_{stop} on leaving the cathode. Thus,

$$eV_{stop} = hf - W$$

Thus Einstein's theory makes a very definite quantitative prediction: if the frequency of the incident light is varied, and V_{stop} plotted as a function of frequency, the slope of the line should be h/e. It is also clear that there is a minimum light frequency for a given metal, that for which the quantum of energy is equal to the work function. Light below that frequency, no matter how bright, will not cause photoemission.

MILLIKAN'S ATTEMPTS TO DISPROVE EINSTEIN'S THEORY

If we accept Einstein's theory, then, this is a completely different way to measure Planck's constant. The American experimental physicist Robert Millikan, who did not accept Einstein's theory, which he saw as an attack on the wave theory of light, worked for ten years, until 1916, on the photoelectric effect. He even devised techniques for scraping clean the metal surfaces inside the vacuum tube.

For all his efforts he found disappointing results: he confirmed Einstein's theory, measuring Planck's constant to within 0.5% by this method. One consolation was that he did get a Nobel prize for this series of experiments.

SPARKS IN GASES

Electric discharges were passed through rarefied gases in the time of Benjamin Franklin - His friend William Watson, an Englishman, remarked: "It was a most delightful spectacle, when the room was darkened, to see the electricity in its passage." The experimental arrangement was to have two small metal plates inside a glass container from which the air was gradually removed by a pump.

The plate connected to the negative side of the electricity supply was called the cathode, that to the positive side the anode. As the pressure was lowered, a spark which fattened into a purplish glowing, writhing snake appeared going from the cathode to the anode. Lowering the pressure further, as Faraday noticed in the 1830's, a dark space opened up near the cathode, now called the Faraday Dark Space.

CATHODE RAYS

The big breakthrough came in the 1850's when Geissler invented a much better (mercury) pump. In 1869, Hittorf found that in a very good vacuum (0.01 mm Hg), the Faraday dark space expanded to fill the whole tube, and the cathode emitted rays that caused the glass to glow where they hit. In 1876, Goldstein called them "cathode rays". In 1879, an Englishman, William Crookes, declared that they must be particles of some sort, and demonstrated that they traveled in straight lines by inserting a Maltese cross in the tube, which cast a sharp shadow on the end of the tube, a demo still in common use 120 years later!

Were these cathode rays waves or particles? Hertz, with his student Lenard, discovered in 1891 that the rays could penetrate a thin aluminum plate, and detected them in the air just outside the tube. Hertz found experimentally that he could not deflect the rays with an electrostatic field applied from outside the tube, and they didn't affect a compass, so he concluded (wrongly) that they must be waves, not particles.

THOMSON DISCOVERS THE ELECTRON

For some odd reason, all the Germans physicists thought the cathode rays were waves, the English thought they were particles (presumably ions of some kind). In 1897, J. J. Thomson announced at the Royal Institution that they were negatively charged particles, and he'd measured e/m, the ratio of charge to mass for a single particle. He achieved this with a specially designed cathode ray tube, the precursor of the tube in a present-day TV set. The cathode was at the far narrow end of the tube, the anode a few centimeters away, with a small hole in its centre.

The rays went from the cathode to the anode, but a narrow beam continued through the hole in the anode to the glass at the other end of the tube, where their impact caused the glass to glow in a small central spot.

Thomson also added *inside the tube* two parallel plates arranged so that the narrow beam of cathode rays went between them on its way to the end of the tube. He found that electrically charging these plates caused the beam to deflect, so the glowing spot moved from the centre of the end of the tube. (This was different from Hertz' result.

Possibly Hertz' externally applied field was shielded by space charge on or near the walls of the tube.) Thomson also added magnets to give a uniform magnetic field in the region between the plates. By adjusting this magnetic field strength to cancel the deflection of the rays caused by the electric field, he was able to measure the speed of the rays, because balancing these two forces

$$eE = evB$$

the charge cancels, and the speed is simply the ratio of electric field strength to magnetic field strength.

Having determined the horizontal speed, Thomson was able to deduce the charge to mass ratio by measuring the deflection in the electric field alone. In this situation, the trajectory of the particle is identical to that of a ball, initially thrown horizontally, veering downwards in a parabolic path. The downward (or upward, depending on the sign of the plates' charge) force is eE, and it operates for a time L/v, where L is the length of the plates. On emerging from the region between the plates, the electron moves in a straight line.

The value of e/m that emerged from this experiment was a complete surprise. It was about 2,000 times the value for the hydrogen ion, which has the largest value of any ion. Once it had been established that the cathode rays were not just uncharged electromagnetic waves, it had been assumed that they were "molecular torrents" - a flood of the same kinds of ions found in electrolysis. Now, apparently, the cathode rays were particles smaller than the smallest atom! This was the first hint that atoms might be made up of smaller entities.

Another important point is that Thomson found this ratio e/m to be independent of the material the cathode was made of. The cathode rays could have been fragments of atoms which were different for different atoms, but evidently this was not the case. Furthermore, he found in 1899 that photoelectrically produced particles had the same e/m, so were probably the same particles.

For the photoelectrically produced particles, Thomson was able to find the charge e, working with his student C. T. R. Wilson, in a cloud chamber experiment. They introduced photoelectrons into a supersaturated vapour. Tiny droplets formed around each free electron. Thomson and Wilson estimated the total number of droplets formed by counting, then found the total charge by collecting the droplets on an electroscope. Their results were off by almost 40%, but close enough to suggest strongly that the particles had the same magnitude charge as the hydrogen ion. This confirmed that the charged particles were much lighter than atoms. (They could have been the same mass as atoms, but carrying a huge charge. This unlikely possibility was now ruled out.)

Thomson had discovered the electron. (Actually, the word electron was first used in 1891 by Stoney, an Irish physicist, to denote the charge on an ion, which he estimated from electrolysis and Avogadro's number.) The emerging picture was that the atom, known of course to be electrically neutral, contained negatively charged particles, the electrons, which could be removed in various ways, leaving a positively charged ion which contained almost all the mass of the original atom.

To put it in his own words, in 1899: "Electrification essentially involves the splitting up of the atom, a part of the mass of the atom getting free and becoming detached from the original atom."(Pais p 86). The word atom comes from the Greek, and means that which cannot be cut up. ("tom" is the same root for cut that appears, for example, in appendectomy). Thomson was the first to talk about splitting atoms.

It had long been known, of course, that in electrolysis of water, for example, by passing an electric current through it, the molecules were somehow split into hydrogen ions, which carried a positive charge and moved to the cathode, where they received a negative electric charge from the battery and appeared as hydrogen atoms combined into hydrogen molecules; and negatively charged oxygen ions, appearing in the same way as oxygen gas at the anode. The new insight was that these ions were atoms having an excess or a deficiency of electrons, and the electrons were particles whose charge and mass was now known.

X-RAYS

But actually in recounting this tale we skipped over an even more important related development in 1895. (Pais p 36) Wilhelm Roentgen was a 50 year old Director of a Physics Institute, a position that included generous family living quarters, including for example, a well stocked wine cellar.

He was checking out the work of Hertz and Lenard on cathode rays, and in particular was interested in Lenard's discovery that the rays penetrated a little way into the air if the Crookes' tube had an aluminum window.

He had covered his Crookes' tube with black cardboard, presumably (my guess) so that he could detect the rays in the air without the distracting glow from the tube itself. He darkened the room to check that no light was getting through, and suddenly (from an unnamed biographer quoted in the Project Physics Text) "noticed a weak light shimmering on a little bench nearby. Highly excited, Roentgen lit a match, and, to his great surprise, discovered that the source of the mysterious light was a little barium platinocyanide screen lying on the bench."

This little screen had been made to detect ultraviolet light, which causes barium platinocyanide to fluoresce. But if the Crookes' tube was producing any ultraviolet light, it wouldn't make it through the cardboard, and the cathode rays themselves only traveled a few centimeters in air. The little screen was a meter away from the tube.

Roentgen concluded that the Crookes' tube must be emitting some new radiation of an unknown type. To reflect his bewilderment, he dubbed the new radiation "x-rays". He couldn't believe it. He checked and rechecked. He told his wife that when he announced it, everybody would say "Roentgen has probably gone crazy".

Before publishing his results, he spent seven weeks investigating the properties of these x-rays. He had already found that ordinary film was sensitive to them. He found lead was a good shield, and, by using it to shield different parts of the Crookes' tube, established that the x-rays originated in the part of the glass that fluoresced where the cathode rays hit it.

He found that the x-rays traveled in straight lines, and (unlike the cathode rays) were not deflected by magnetic fields. He found they passed through flesh almost unimpeded, but bone cast a shadow. By having his wife place her hand between the point source of x-rays on the Crookes' tube and some unexposed film in a box, then developing the film, he took a picture of the bones in her hand. On January 1, 1896 he announced his findings, complete with the bone picture, to many of his colleagues, and remarked to his wife "Now the fun begins".

The x-ray story is a good antidote to the myth that new scientific discoveries are developed and utilized much more rapidly than they used to be. On the 25th of January, a young research student in Cambridge, Ernest Rutherford, wrote of his Professor:

"The Professor, of course, is trying to find out the real cause and nature of the waves, and the great object is to find the theory of matter before anyone else, for nearly every Professor in Europe is now on the warpath."

Meanwhile, in the USA, the New York Times reported on the 9th of February: "the wizard of New Jersey (Edison) will try to photograph the skeleton of a human head next week." (He didn't succeed.) Already in 1896 several hospitals had x-ray facilities, and x-ray photographs were ruled as acceptable evidence in courts in France, England and the USA.

Before the end of the year, the dangers became apparent. A Columbia professor who had been giving demonstrations at Bloomingdale's suffered severe skin damage. Over the next five years, many doctors and patients suffered horribly, and patients began to sue successfully. By 1903, lead-impregnated rubber shielding devices were being used.

WHAT ARE X-RAYS?

All the activity described above did little to clarify the actual nature of x-rays. Roentgen himself found they were undeflected by a magnetic field, so were not charged particles like the cathode rays. On the other hand, they didn't exhibit any diffraction phenomena, so did not seem to be waves. Roentgen, and independently Thomson, found that the x-rays were ionizing radiation - as they passed through air, ions were created. A gold leaf electroscope exposed to the x-rays would lose its charge, as the newly created ions were attracted to the charged leaves.

In 1899, Haga and Wind noticed a slight broadening of an x-ray beam after it passed through a slit a few thousandths of a millimeter wide. This could be from diffraction if the wavelength were of order 10^{-10} meters. This problem was not resolved conclusively until 1912, when Laue made the observation that since the wavelength of x-rays was apparently similar to the distances between planes of atoms in a crystal, perhaps a crystal would act as a diffraction grating for x-rays. This turned out to be correct, and in fact is now the standard way of finding crystal structure. It was concluded, then, that x-rays were "ultra- ultra violet light", as one French physicist had put it in 1896.

Once the usefulness of x-rays was established, techniques for producing them evolved rapidly. It was found that they were produced far more copiously if the cathode rays impinged on a piece of heavy metal, such as Molybdenum, rather than glass. The physical picture of x-ray production was that the electrons radiated as they suddenly decelerated on hitting the target, unloading their kinetic energy as radiation (plus some heating of the target). This deceleration radiation is called bremsstrahlung in German, and this word is sometimes used to describe it.

X-RAYS COME IN QUANTA

Since crystals act as diffraction gratings for x-rays, it is possible to map out a spectrum of the x-rays. (Of course, this was not done until after 1912). The most important finding on doing this was that there was a maximum frequency (minimum wavelength) at which x-ray production stopped, and this maximum frequency depended on the voltage applied to the cathode ray tube in a simple way:

$$hf_{max} = eV$$

The explanation is of course very simple - each electron in the cathode rays gains a kinetic energy of eV on accelerating down the tube before crashing into the target. Therefore, this is the most energy that can be emitted as it loses kinetic energy in the crash. If this all goes into one photon, the frequency will be f_{max}. So the cutoff in the x-ray spectrum provides another way to measure Planck's constant.

MORE RAYS

To return now to 1896, at first nobody had any idea how the x-rays were being generated, but Roentgen had clearly established that they came from the bit of Crookes' tube glass that was fluorescing. A French physicist, Henri Becquerel, attended a meeting of the French Academy of Sciences on January 20, 1896, *three weeks* after Roentgen's first announcement. Two French physicians were already showing hand x-rays.

It occurred to Becquerel that any substance that fluoresced intensely might also emit x-rays. It happened that fifteen years previously he had worked with a substance that fluoresced strongly when exposed to sunlight, so he wondered if it might also be emitting x-rays, previously undetected.

To find out, he placed a sheet of it in the sun, lying on top of a cardboard box containing unexposed film with a small metal object above it. After a day's exposure to the sunlight, which caused fluorescence, he took out and developed the film. Sure enough, it was exposed except in the shadow of the metal object, which appeared in silhouette.

He decided to check, and set up the same arrangement again, with a small metal cross above the film. Unfortunately, however, it continued cloudy, so he kept the package in a closet waiting for the sun to reappear. After several days Becquerel grew impatient, and, possibly not having enough to do, decided to develop the film, perhaps to check if some light had leaked in. To his utter astonishment, he found the silhouette of the cross on the film just as distinctly as before!

Question for the reader: before reading on, can you figure out any possible explanation for this surprising discovery?

Evidently, the recent exposure to sunlight had not been a crucial element in the production of the mysterious radiation that had exposed the film. Becquerel theorized that maybe the substance continued to manufacture rays long after the fluorescence died away. However, it soon became clear that in fact the fluorescence itself was irrelevant.

It happened that the fluorescent substance Becquerel used for his experiment contained uranium. After extensive experimenting with various substances, he concluded by May that the radiation came from any substances containing uranium, whether they fluoresced or not.

He found the intensity of the radiation increased with the amount of uranium present, and did not appear to change in intensity with time, or temperature, or chemical action. Like x-rays, the radiation was ionizing. If a piece of uranium is held near an electroscope, the electroscope discharges. Unlike x-rays, though, these rays couldn't be turned off.

MARIE CURIE INVESTIGATES BECQUEREL'S RAYS

In contrast to all the hoopla over x-rays, Becquerel's rays

had little public impact, at least at the beginning. Fortunately, they did attract the attention of two very talented young researchers, Marie Curie in Paris and Ernest Rutherford in Cambridge. Marie worked with her husband Pierre, carefully measuring the amount of radiation from different amounts of uranium compounds. They measured the radiation by using two parallel metal plates differing in potential by 100 volts, connected by a sensitive electrometer (basically an infinite resistance voltmeter, a calibrated electroscope).

A thin layer of the uranium compound under examination was then put on the top plate, and the rate at which charge was lost was measured - the ionizing radiation caused the charge leakage. Marie concluded that the radiation intensity was just proportional to the number of uranium atoms present, so the radiation originated inside the uranium atom. This conclusion was harder to reach than it sounds, because some fraction of the radiation is absorbed in the material itself. This is why she used a thin layer, rather than a piece of the material under investigation.

The mineral pitchblende, rich in uranium oxide, was found to be more radioactive than pure uranium metal. Marie concluded that it must contain some other element that was more radioactive than uranium. In 1898, the Curies chemically separated out a new element they called polonium, after Marie's country of birth, Poland.

They measured its radioactivity as four hundred times more intense than that from pure uranium. Six months later, they chemically separated out from pitchblende another radioactive substance more than one *million* times more radioactive than uranium. They called it radium. One gram of radium gives out enough energy to raise one gram of water from just above freezing to boiling in one hour.

Where this energy could be coming from was a complete mystery. It was hard to imagine how the energy could be stores in the material - in a few days, radium gave off as much energy as the most potent chemical fuel, yet it kept on radiating year after year.

One popular theory at the time was that space was full of waves of energy, and almost all materials were completely transparent to this sea of radiation, but radium, polonium and uranium absorbed some of this energy and re-emitted it.

ERNEST RUTHERFORD INVESTIGATES BECQUEREL'S RAYS

Rutherford worked as a student with J. J. Thomson from 1896, studying the ionization of gases by x-rays in a quantitative fashion. When Becquerel announced his discovery of new rays, it was natural for Rutherford to investigate them as well. He began work in Cambridge, but then accepted a job at McGill University in Montreal. He needed some money because he wanted to get married. He wrote to his fiancee: "I am expected to do a lot of original work and to form a research school to knock the shine out of the Yankees!"

He carefully studied absorption of the Becquerel rays, and found that one component, which he called the alpha-rays, could be absorbed by a sheet of writing paper, or a few centimeters of air. In fact, Becquerel had not detected these alphas, because they were absorbed by the box containing the film. A second component, the beta rays, Rutherford found to be one hundred times more penetrating. He published this result in 1899. In 1900, Rutherford went home to New Zealand and got married. French physicists, meanwhile, kept up the pace of discovery. Villard found that radium emitted some far more penetrating radiation, which he christened gamma rays. These rays could penetrate several feet of concrete.

IDENTIFYING THE BETA RAYS

It was clear by 1900 from their deflection by a magnetic field that the beta rays were negatively charged particles. In that year, Becquerel found e/m for beta rays to be quite close to that for cathode rays, suggesting that they also were electrons.

One difference between the beta rays and cathode rays was that the beta rays could be much faster - up to 95% of the speed of light. By 1902, a German physicist, Kaufman was writing: "for small velocities,.. [e/m for] Becquerel rays ... fits within experimental error with the value found for cathode rays." (my italics). The interesting historical point here is that physicists already expected that mass might change with speed. There were complicated (and wrong) electromagnetic theories about how the mass might vary, beginning with the idea that the kinetic energy of a charge moving through the ether included energy from the ether flowing around it - an idea based on earlier analyses of the motion of a sphere through a real fluid.

Lorentz contraction effects were added to give yet more complicated formulas. However, in 1908, a German experimenter, Bucherer, claimed that the best fit to the *e/m* data was given by Einstein's much simpler recent formula for mass variation with speed. These experiments were difficult to do, and Bucherer's results were not universally accepted until about 1916, by which time others had found the same results.

IDENTIFYING THE ALPHA RAYS

The less penetrating alpha radiation proved much more difficult to identify. At first the alphas were thought to be electrically neutral, because there was no observed deflection as they shot through a magnetic field. It became clear later that this lack of deflection was because the alphas were much more massive than the betas (electrons). In 1903, in a more careful experiment, Rutherford found the alphas were in fact deflected slightly, and in the opposite direction to the electrons, so they were positively charged. By 1905, he had found e/m, and concluded that if the charge on an alpha was the same as that on a hydrogen ion, the mass of the alpha was approximately twice that of the hydrogen atom. In 1908, he finally established that the alphas were helium atoms with two electrons missing, carrying charge 2e, and having mass four times that of the hydrogen atom.

Identifying the Gamma Rays

The very penetrating gamma rays, discovered in 1900, also took years to identify. They were not deflected by a magnetic field. Rutherford, in 1902, thought (wrongly) they might be extremely fast beta particles. It gradually became clear that they were akin to x-rays, but with much shorter wavelength. This was not settled until 1914, when Rutherford observed them to be reflected from crystal surfaces.

Chapter 8

Special Relativity

FRAMES OF REFERENCE AND NEWTON'S LAWS

The cornerstone of the theory of special relativity is the Principle of Relativity: The Laws of Physics are the same in all inertial frames of reference.

We shall see that many surprising consequences follow from this innocuous looking statement.

Let us first, however, briefly review Newton's mechanics in terms of frames of reference.

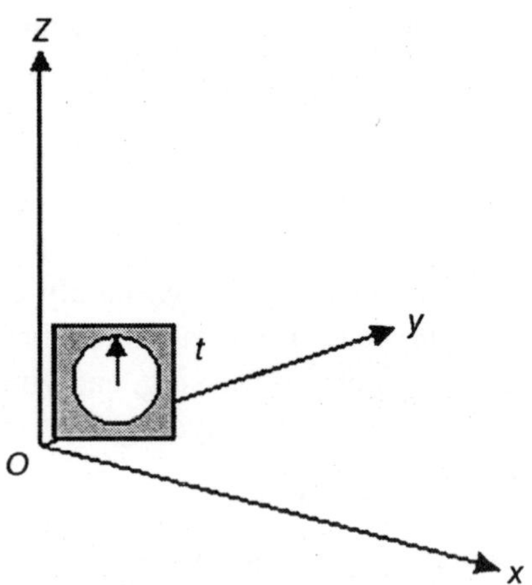

Fig. A frame of Reference

A "frame of reference" is just a set of coordinates: something you use to measure the things that matter in Newtonian problems, that is to say, positions and velocities, so we also need a clock.

A point in space is specified by its three coordinates (x, y,

z) and an "event" like, say, a little explosion, by a place and time: (x, y, z, t).

An inertial frame is defined as one in which Newton's law of inertia holds—that is, any body which isn't being acted on by an outside force stays at rest if it is initially at rest, or continues to move at a constant velocity if that's what it was doing to begin with. An example of a *non*-inertial frame is a rotating frame, such as a carousel.

The "laws of physics" we shall consider first are those of Newtonian mechanics, as expressed by Newton's Laws of Motion, with gravitational forces and also contact forces from objects pushing against each other. For example, knowing the universal gravitational constant from experiment (and the masses involved), it is possible from Newton's Second Law,

$$\text{force} = \text{mass} \times \text{acceleration},$$

to predict future planetary motions with great accuracy.

Suppose we know from experiment that these laws of mechanics are true in one frame of reference. How do they look in another frame, moving with respect to the first frame? To find out, we have to figure out how to get from position, velocity and acceleration in one frame to the corresponding quantities in the second frame.

Obviously, the two frames must have a constant relative velocity, otherwise the law of inertia won't hold in both of them. Let's choose the coordinates so that this velocity is along the x-axis of both of them.

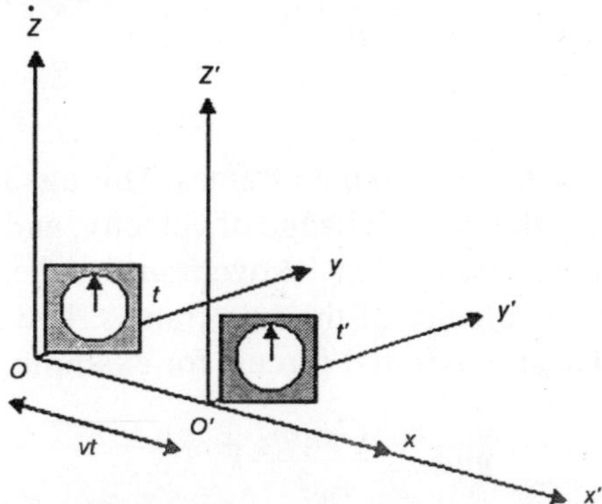

Fig. Two frames of Reference Relatively Displaced along the x-axis

Notice we also throw in a clock with each frame.

Suppose S' is proceeding relative to S at speed v along the x-axis. For convenience, let us label the moment when O' passes O as the zero point of timekeeping. Now what are the coordinates of the event (x, y, z, t) in S'? It's easy to see $t' = t$—we synchronized the clocks when O' passed O. Also, evidently, $y' = y$ and $z' = z$, from the figure. We can also see that $x = x' + vt$. Thus (x, y, z, t) in S corresponds to (x', y', z', t') in S', where

$$x' = x - vt$$
$$y' = y$$
$$z' = z$$
$$t' = t.$$

That's how positions transform; these are known as the Galilean transformations. What about *velocities* ? The velocity in S' in the x' direction

$$u'_x = \frac{dx'}{dt'} = \frac{dx'}{dt} = \frac{d}{dt}(x - vt) = \frac{dx}{dt} - v - u_x - v$$

This is obvious anyway: it's just the addition of velocities formula

$$u_x = u'_x + v.$$

How does *acceleration* transform?

$$\frac{du'_x}{dt'} = \frac{du'_x}{dt} = \frac{d}{dt}(u_x - v) = \frac{du_x}{dt}$$

since v is constant.

That is to say, $a'_x = a_x$

the acceleration is the same in both frames. This again is obvious—the acceleration is the rate of change of velocity, and the velocities of the same particle measured in the two frames differ by a *constant* factor-the relative velocity of the two frames. If we now look at the motion under gravitational forces, for example,

$$m_1\vec{a} = \frac{Gm_1m_2}{r^2}\hat{\vec{r}}$$

we get the same law on going to another inertial frame because every term in the above equation stays the same Note that $m\vec{a}$ is the rate of change of momentum—this is the same in both frames. So, in a collision, say, if total momentum is conserved in one frame (the sum of individual rates of change of momentum is zero) the same is true in *all* inertial frames.

THE SPEED OF LIGHT

EARLY IDEAS ABOUT LIGHT PROPAGATION

Attempts to measure the speed of light played an important part in the development of the theory of special relativity, and, indeed, the speed of light is central to the theory. The first recorded discussion of the speed of light (I think) is in Aristotle, where he quotes Empedocles as saying the light from the sun must take some time to reach the earth, but Aristotle himself apparently disagrees, and even Descartes thought that light traveled instantaneously.

SIMP. Everyday experience shows that the propagation of light is instantaneous; for when we see a piece of artillery fired at great distance, the flash reaches our eyes without lapse of time; but the sound reaches the ear only after a noticeable interval.

Of course, Galileo points out that in fact nothing about the speed of light can be deduced from this observation, except that light moves faster than sound. He then goes on to suggest a possible way to measure the speed of light. The idea is to have two people far away from each other, with covered lanterns.

One uncovers his lantern, then the other immediately uncovers his on seeing the light from the first. This routine is to be practiced with the two close together, so they will get used to the reaction times involved, then they are to do it two or three miles apart, or even further using telescopes, to see if the time interval is perceptibly lengthened. Galileo claims he actually tried the experiment at distances less than a mile, and couldn't detect a time lag. From this one can certainly deduce that light travels at least ten times faster than sound.

MEASURING THE SPEED OF LIGHT WITH JUPITER'S MOONS

The first real measurement of the speed of light came about

half a century later, in 1676, by a Danish astronomer, Ole Rφmer, working at the Paris Observatory. He had made a systematic study of Io, one of the moons of Jupiter, which was eclipsed by Jupiter at regular intervals, as Io went around Jupiter in a circular orbit at a steady rate. Actually, Rφmer found, for several months the eclipses lagged more and more behind the expected time, until they were running about eight minutes late, then they began to pick up again, and in fact after about six months were running eight minutes early.

The cycle then repeated itself. Rφmer realized the significance of the time involved-just over one year. This time period had nothing to do with Io, but was the time between successive closest approaches of earth in its orbit to Jupiter. The eclipses were furthest behind the predicted times when the earth was furthest from Jupiter.

The natural explanation was that the light from Io (actually reflected sunlight, of course) took time to reach the earth, and took the longest time when the earth was furthest away. From his observations, Rφmer concluded that light took about twenty-two minutes to cross the earth's orbit.

This was something of an overestimate, and a few years later Newton wrote in the *Principia*: "For it is now certain from the phenomena of Jupiter's satellites, confirmed by the observations of different astronomers, that light is propagated in succession and requires about seven or eight minutes to travel from the sun to the earth." This is essentially the correct value. Of course, to find the speed of light it was also necessary to know the distance from the earth to the sun.

During the 1670's, attempts were made to measure the parallax of Mars, that is, how far it shifted against the background of distant stars when viewed simultaneously from two different places on earth at the same time. This (very slight) shift could be used to find the distance of Mars from earth, and hence the distance to the sun, since all *relative* distances in the solar system had been established by observation and geometrical analysis. According to Crowe, they concluded that the distance to the sun was between 40 and 90 million miles.

Measurements presumably converged on the correct value of

about 93 million miles soon after that, because it appears Rφmer (or perhaps Huygens, using Rφmer's data a short time later) used the correct value for the distance, since the speed of light was calculated to be 125,000 miles per second, about three-quarters of the correct value of 186,300 miles per second. This error is fully accounted for by taking the time light needs to cross the earth's orbit to be twenty-two minutes (as Rφmer did) instead of the correct value of sixteen minutes.

STARLIGHT AND RAIN

The next substantial improvement in measuring the speed of light took place in 1728, in England. An astronomer James Bradley, sailing on the Thames with some friends, noticed that the little pennant on top of the mast changed position each time the boat put about, even though the wind was steady. He thought of the boat as the earth in orbit, the wind as starlight coming from some distant star, and reasoned that the apparent direction the starlight was "blowing" in would depend on the way the earth was moving.

Another possible analogy is to imagine the starlight as a steady downpour of rain on a windless day, and to think of yourself as walking around a circular path at a steady pace. The apparent direction of the incoming rain will not be vertically downwards-more will hit your front than your back. In fact, if the rain is falling at, say, 15 mph, and you are walking at 3 mph, to you as observer the rain will be coming down at a slant so that it has a vertical speed of 15 mph, and a horizontal speed towards you of 3 mph. Whether it is slanting down from the north or east or whatever at any given time depends on where you are on the circular path at that moment.

Bradley reasoned that the apparent direction of incoming starlight must vary in just this way, but the angular change would be a lot less dramatic. The earth's speed in orbit is about 18 miles per second, he knew from Rφmer's work that light went at about 10,000 times that speed. That meant that the angular variation in apparent incoming direction of starlight was about the magnitude of the small angle in a right-angled triangle with one side 10,000 times longer than the other, about one two-hundredth of a degree.

Notice this would have been just at the limits of Tycho's measurements, but the advent of the telescope, and general improvements in engineering, meant this small angle was quite accurately measurable by Bradley's time, and he found the velocity of light to be 185,000 miles per second, with an accuracy of about one percent.

FAST FLICKERING LANTERNS

The problem is, all these astronomical techniques do not have the appeal of Galileo's idea of two guys with lanterns. It would be reassuring to measure the speed of a beam of light between two points on the ground, rather than making somewhat indirect deductions based on apparent slight variations in the positions of stars. We can see, though, that if the two lanterns are ten miles apart, the time lag is of order one-ten thousandth of a second, and it is difficult to see how to arrange that.

This technical problem was solved in France about 1850 by two rivals, Fizeau and Foucault, using slightly different techniques. In Fizeau's apparatus, a beam of light shone between the teeth of a rapidly rotating toothed wheel, so the "lantern" was constantly being covered and uncovered. Instead of a second lantern far away, Fizeau simply had a mirror, reflecting the beam back, where it passed a second time between the teeth of the wheel. The idea was, the blip of light that went out through one gap between teeth would only make it back through the same gap if the teeth had not had time to move over significantly during the round trip time to the far away mirror.

It was not difficult to make a wheel with a hundred teeth, and to rotate it hundreds of times a second, so the time for a tooth to move over could be arranged to be a fraction of one ten thousandth of a second.

The method worked. Foucault's method was based on the same general idea, but instead of a toothed wheel, he shone the beam on to a rotating mirror. At one point in the mirror's rotation, the reflected beam fell on a distant mirror, which reflected it right back to the rotating mirror, which meanwhile had turned through a small angle.

After this second reflection from the rotating mirror, the

position of the beam was carefully measured. This made it possible to figure out how far the mirror had turned during the time it took the light to make the round trip to the distant mirror, and since the rate of rotation of the mirror was known, the speed of light could be figured out. These techniques gave the speed of light with an accuracy of about 1,000 miles per second.

ALBERT ABRAHAM MICHELSON

Albert Michelson was born in 1852 in Strzelno, Poland. His father Samuel was a Jewish merchant, not a very safe thing to be at the time. Purges of Jews were frequent in the neighboring towns and villages. They decided to leave town. Albert's fourth birthday was celebrated in Murphy's Camp, Calaveras County, about fifty miles south east of Sacramento, a place where five million dollars worth of gold dust was taken from one four acre lot. Samuel prospered selling supplies to the miners.

When the gold ran out, the Michelsons moved to Virginia City, Nevada, on the Comstock lode, a silver mining town. Albert went to high school in San Francisco. In 1869, his father spotted an announcement in the local paper that Congressman Fitch would be appointing a candidate to the Naval Academy in Annapolis, and inviting applications.

Albert applied but did not get the appointment, which went instead to the son of a civil war veteran. However, Albert knew that President Grant would also be appointing ten candidates himself, so he went east on the just opened continental railroad to try his luck. Unknown to Michelson, Congressman Fitch wrote directly to Grant on his behalf, saying this would really help get the Nevada Jews into the Republican party.

This argument proved persuasive. In fact, by the time Michelson met with Grant, all ten scholarships had been awarded, but the President somehow came up with another one. Of the incoming class of ninety-two, four years later twenty-nine graduated. Michelson placed first in optics, but twenty-fifth in seamanship.

The Superintendent of the Academy, Rear Admiral Worden, who had commanded the Monitor in its victory over the Merrimac, told Michelson: "If in the future you'd give less attention to those scientific things and more to your naval gunnery, there might come a time when you would know enough to be of some service to your country."

SAILING THE SILENT SEAS: GALILEAN RELATIVITY

Shortly after graduation, Michelson was ordered aboard the USS Monongahela, a sailing ship, for a voyage through the Caribbean and down to Rio. According to the biography of Michelson written by his daughter (The Master of Light, by Dorothy Michelson Livingston, Chicago, 1973) he thought a lot as the ship glided across the quiet Caribbean about whether one could decide in a closed room inside the ship whether or not the vessel was moving. In fact, his daughter quotes a famous passage from Galileo on just this point:

Shut yourself up with some friend in the largest room below decks of some large ship and there procure gnats, flies, and other such small winged creatures. Also get a great tub full of water and within it put certain fishes; let also a certain bottle be hung up, which drop by drop lets forth its water into another narrow-necked bottle placed underneath.

Then, the ship lying still, observe how those small winged animals fly with like velocity towards all parts of the room; how the fish swim indifferently towards all sides; and how the distilling drops all fall into the bottle placed underneath.

And casting anything toward your friend, you need not throw it with more force one way than another, provided the distances be equal; and leaping with your legs together, you will reach as far one way as another. Having observed all these particulars, though no man doubts that, so long as the vessel stands still, they ought to take place in this manner, make the ship move with what velocity you please, so long as the motion is uniform and not fluctuating this way and that. You will not be able to discern the least alteration in all the forenamed effects, nor can you gather by any of them whether the ship moves or stands still....in throwing something to your friend you do not need to throw harder if he is towards the front of the ship from you... the drops from the upper bottle still fall into the lower bottle even though the ship may have moved many feet while the drop is in the air...

Of this correspondence of effects the cause is that the ship's motion is common to all the things contained in it and to the air also; I mean if those things be shut up in the room; but in case those things were above the deck in the open air, and not obliged to follow the course of the ship, differences would be observed,... smoke would stay behind.

MICHELSON MEASURES THE SPEED OF LIGHT

On returning to Annapolis from the cruise, Michelson was commissioned Ensign, and in 1875 became an instructor in physics and chemistry at the Naval Academy, under Lieutenant Commander William Sampson. Michelson met Mrs. Sampson's niece, Margaret Heminway, daughter of a very successful Wall Street tycoon, who had built himself a granite castle in New Rochelle, NY. Michelson married Margaret in an Episcopal service in New Rochelle in 1877. At work, lecture demonstrations had just been introduced at Annapolis. Sampson suggested that it would be a good demonstration to measure the speed of light by Foucault's method. Michelson soon realized, on putting together the apparatus, that he could redesign it for much greater accuracy, but that would need money well beyond that available in the teaching demonstration budget.

He went and talked with his father in law, who agreed to put up $2,000. Instead of Foucault's 60 feet to the far mirror, Michelson had about 2,000 feet along the bank of the Severn, a distance he measured to one tenth of an inch. He invested in very high quality lenses and mirrors to focus and reflect the beam. His final result was 186,355 miles per second, with possible error of 30 miles per second or so. This was twenty times more accurate than Foucault, made the New York Times, and Michelson was famous while still in his twenties. In fact, this was accepted as the most accurate measurement of the speed of light for the next forty years, at which point Michelson measured it again.

THE MICHELSON-MORLEY EXPERIMENT

THE NATURE OF LIGHT

Michelson's efforts in 1879, the speed of light was known to be 186,350 miles per second with a likely error of around 30 miles per second. This measurement, made by timing a flash of light travelling between mirrors in

Annapolis, agreed well with less direct measurements based on astronomical observations. Still, this did not really clarify the *nature* of light. Two hundred years earlier, Newton had suggested that light consists of tiny *particles* generated in a hot object, which spray out at very high speed, bounce off other objects, and are detected by our eyes. Newton's arch-enemy Robert Hooke, on the other hand, thought that light must be a kind of *wave motion*, like sound. To appreciate his point of view, let us briefly review the nature of sound.

THE WAVELIKE NATURE OF SOUND

Actually, *sound* was already quite well understood by the ancient Greeks. The essential point they had realized is that sound is generated by a vibrating material object, such as a bell, a string or a drumhead. Their explanation was that the vibrating drumhead, for example, alternately pushes and pulls on the air directly above it, sending out waves of compression and decompression (known as rarefaction), like the expanding circles of ripples from a disturbance on the surface of a pond.

On reaching the ear, these waves push and pull on the eardrum with the same frequency (that is to say, the same number of pushes per second) as the original source was vibrating at, and nerves transmit from the ear to the brain both the intensity (loudness) and frequency (pitch) of the sound.

There are a couple of special properties of sound waves (actually any waves) worth mentioning at this point. The first is called *interference*. This is most simply demonstrated with water waves. If you put two fingers in a tub of water, just touching the surface a foot or so apart, and vibrate them at the same rate to get two expanding circles of ripples, you will notice that where the ripples overlap there are quite complicated patterns of waves formed.

The essential point is that at those places where the wave-crests from the two sources arrive at the same time, the waves will work together and the water will be very disturbed, but at points where the crest from one source arrives at the same time as the wave trough from the other source, the waves will cancel each other out, and the water will hardly move. You can hear this effect for sound waves by playing a constant note through stereo speakers. As you move around a room, you will hear quite large variations in the intensity of sound. Of course, reflections from walls complicate the pattern.

This large variation in volume is *not* very noticeable when the stereo is playing music, because music is made up of many frequencies, and they change all the time. The different frequencies, or notes, have their quiet spots in the room in different places. The other point that should be mentioned is that high frequency tweeter-like sound is much more *directional* than low frequency woofer-like sound. It really doesn't matter where in the room you put a low-frequency woofer—the sound seems to be all around you anyway.

On the other hand, it is quite difficult to get a speaker to spread the high notes in all directions. If you listen to a cheap speaker, the high notes are loudest if the speaker is pointing right at you. A lot of effort has gone into designing tweeters, which are small speakers especially designed to broadcast high notes over a wide angle of directions.

IS LIGHT A WAVE?

Bearing in mind the above minireview of the properties of waves, let us now reconsider the question of whether light consists of a stream of particles or is some kind of wave. The strongest argument for a particle picture is that light travels in straight lines. You can hear around a corner, at least to some extent, but you certainly can't see.

Furthermore, no wave-like interference effects are very evident for light. Finally, it was long known, as we have mentioned, that sound waves were compressional waves in air. If light is a wave, just what is waving? It clearly isn't just air, because light reaches us from the sun, and indeed from stars, and we know the air doesn't stretch that far, or the planets would long ago have been slowed down by air resistance.

Despite all these objections, it was established around 1800 that light *is* in fact some kind of wave. The reason this fact had gone undetected for so long was that the wavelength is *really* short, about one fifty-thousandth of an inch. In contrast, the shortest wavelength sound detectable by humans has a wavelength of about half an inch.

The fact that light travels in straight lines is in accord with observations on sound that the higher the frequency (and shorter the wavelength) the greater the tendency to go in straight lines. Similarly, the interference patterns mentioned above for sound waves or ripples on a pond vary over distances of the same sort of size as the wavelengths involved.

Patterns like that would not normally be noticeable for light because they would be on such a tiny scale. In fact, it turns out, there *are* ways to see interference effects with light. A familiar example is the many colors often visible in a soap bubble. These come about because looking at a soap bubble you see light reflected from both sides of a very thin film of water—a thickness that turns out to be comparable to the wavelength of light.

The light reflected from the lower layer has to go a little further to reach your eye, so that light wave must wave an extra time or two before getting to your eye compared with the light reflected from the top layer. What you actually *see* is the *sum* of the light reflected from the top layer and that reflected from the bottom layer.

Thinking of this now as the sum of two sets of waves, the light will be bright if the crests of the two waves arrive together, dim if the *crests* of waves reflected from the top layer arrive simultaneously with the *troughs* of waves reflected from the bottom layer. Which of these two possibilities actually occurs for reflection from a particular bit of the soap film depends on just how much further the light reflected from the lower surface has to travel to reach your eye compared with light from the upper surface, and that depends on the angle of reflection and the thickness of the film.

Suppose now we shine white light on the bubble. White light is made up of all the colors of the rainbow, and these different colors have different wavelengths, so we see colors reflected, because for a particular film, at a particular angle, some colors will be reflected brightly (the crests will arrive together), some dimly, and we will see the ones that win.

IF LIGHT *IS* A WAVE, WHAT IS WAVING?

Having established that light is a wave, though, we still haven't answered one of the major objections raised above. Just what is waving? We discussed sound waves as waves of compression in air.

Actually, that is only one case—sound will also travel through liquids, like water, and solids, like a steel bar. It is found experimentally that, other things being equal, sound travels faster through a medium that is harder to compress: the material just springs back faster and the wave moves through more rapidly. For media of equal springiness, the sound goes faster through the less heavy medium, essentially because the same amount of springiness can push things along faster in a lighter material. So when a sound wave passes, the material—air, water or solid—waves as it goes through.

Taking this as a hint, it was natural to suppose that light must be just waves in some mysterious material, which was called the *aether*, surrounding and permeating everything. This aether must also fill all of space, out to the stars, because we can see them, so the medium must be there to carry the light. (We could never *hear* an explosion on the moon, however loud, because there is no air to carry the sound to us.)

Let us think a bit about what properties this aether must have. Since light travels so fast, it must be very light, and very hard to compress. Yet, as mentioned above, it must allow solid bodies to pass through it freely, without aether resistance, or the planets would be slowing down. Thus we can picture it as a kind of ghostly wind blowing through the earth. But how can we prove any of this? Can we detect it?

DETECTING THE AETHER WIND: THE MICHELSON-MORLEY EXPERIMENT

Detecting the aether wind was the next challenge Michelson set himself after his triumph in measuring the speed of light so accurately. Naturally, something that allows solid bodies to pass through it freely is a little hard to get a grip on. But Michelson realized that, just as the speed of sound is relative to the air, so the speed of light must be relative to the aether.

This must mean, if you could measure the speed of light accurately enough, you could measure the speed of light travelling upwind, and compare it with the speed of light travelling downwind, and the difference of the two measurements should be twice the windspeed.

Unfortunately, it wasn't that easy. All the recent accurate measurements had used light travelling to a distant mirror and coming back, so if there was an aether wind along the direction between the mirrors, it would have opposite effects on the two parts of the measurement, leaving a very small overall effect. There was no technically feasible way to do a one-way determination of the speed of light.

At this point, Michelson had a very clever idea for detecting the aether wind. As he explained to his children, it was based on the following puzzle:

Suppose we have a river of width w (say, 100 feet), and two swimmers who both swim at the same speed v feet per second (say, 5 feet per second). The river is flowing at a steady rate, say 3 feet per second. The swimmers race in the following way: they both start at the same point on one bank. One swims directly across the river to the closest point on the opposite bank, then turns around and swims back. The other stays on one side of the river, swimming upstream a distance (measured along the bank) exactly equal to the width of the river, then swims back to the start. Who wins?

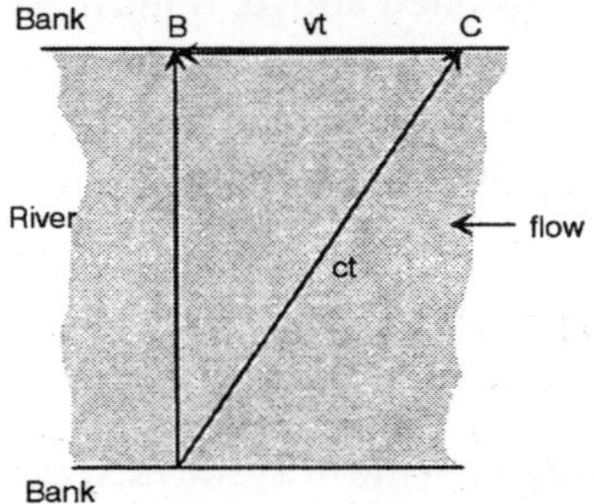

Fig. In time *t*, the Swimmer has moved *ct* Relative to the water, and been Carried Downstream a Distance *vt*.

Let's consider first the swimmer going upstream and back. Going 100 feet upstream, the speed relative to the bank is only 2 feet per second, so that takes 50 seconds. Coming back, the speed is 8 feet per second, so it takes 12.5 seconds, for a total time of 62.5 seconds.

The swimmer going across the flow is trickier. It won't do simply to aim directly for the opposite bank: the flow will carry the swimmer downstream. To succeed in going directly across, the swimmer must actually aim upstream at the correct angle (of course, a real swimmer would do this automatically). Thus, the swimmer is going at 5 feet per second, at an angle, relative to the river, and being carried downstream at a rate of 3 feet per second.

If the angle is correctly chosen so that the net movement is directly across, in one second the swimmer must have moved *four feet* across: the distances covered in one second will form a 3,4,5 triangle. So, at a crossing rate of 4 feet per second, the swimmer gets across in 25 seconds, and back in the same time, for a total time of 50 seconds. The cross-stream swimmer wins. This turns out to true whatever their swimming speed. (Of course, the race is only possible if they can swim faster than the current!)

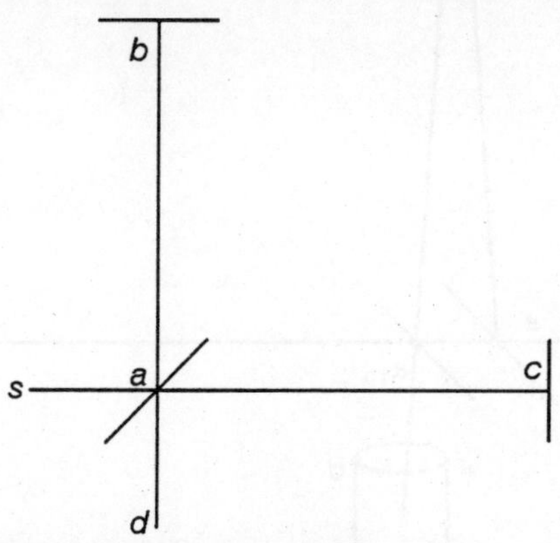

Fig. This Diagram is from the Original paper. The source of light is at *s*, the 45 degree line is the half-silvered mirror, *b* and *c* are mirrors and d the observer.

Michelson's great idea was to construct an exactly similar race for pulses of light, with the aether wind playing the part of the river. The scheme of the experiment is as follows: a pulse of light is directed at an angle of 45 degrees at a half-silvered, half transparent mirror, so that half the pulse goes on through the glass, half is reflected. These two half-pulses are the two swimmers. They both go on to distant mirrors which reflect them back to the half-silvered mirror.

At this point, they are again half reflected and half transmitted, but a telescope is placed behind the half-silvered mirror as shown in the figure so that half of each half-pulse will arrive in this telescope. Now, if there is an aether wind blowing, someone looking through the telescope should see the halves of the two half-pulses to arrive at slightly different times, since one would have gone more upstream and back, one more across stream in general. To maximize the effect, the whole apparatus, including the distant mirrors, was placed on a large turntable so it could be swung around.

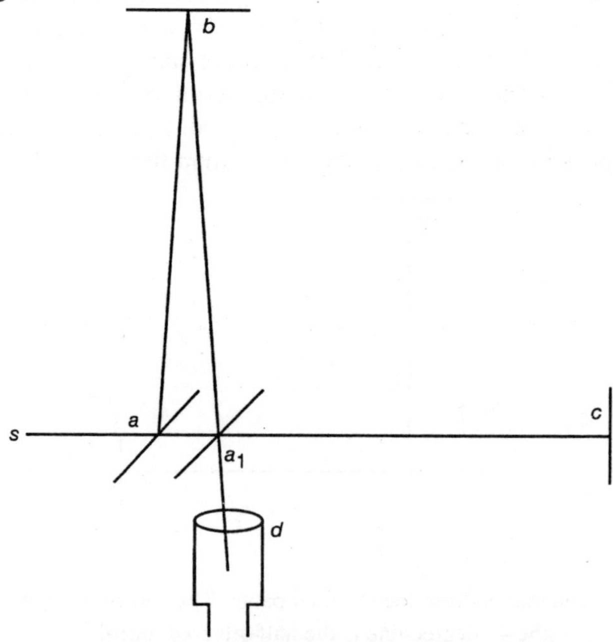

Fig. This is also from the Original paper, and shows the Expected path of light relative to the aether with an aether wind blowing.

Let us think about what kind of time delay we expect to find between the arrival of the two half-pulses of light. Taking the speed of light to be c miles per second relative to the aether, and the aether to be flowing at v miles per second through the laboratory, to go a distance w miles upstream will take $w/(c-v)$ seconds, then to come back will take $w/(c+v)$ seconds.

The total roundtrip time upstream and downstream is the sum of these, which works out to be $2wc/(c^2-v^2)$, which can also be written $(2w/c)\times 1/(1-v^2/c^2)$. Now, we can safely assume the speed of the aether is much less than the speed of light, otherwise it would have been noticed long ago, for example in timing of eclipses of Jupiter's satellites.

This means v^2/c^2 is a very small number, and we can use some handy mathematical facts to make the algebra a bit easier. First, if x is very small compared to 1, $1/(1-x)$ is very close to $1+x$. (You can check it with your calculator.) Another fact we shall need in a minute is that for small x, the square root of $1+x$ is very close to $1+x/2$.

Putting all this together,

$$\text{upstream-downstream roun dtrip time} \cong \frac{2w}{c}\times\left(1+\frac{v^2}{c^2}\right).$$

Now, what about the cross-stream time? The actual cross-stream speed must be figured out as in the example above using a right-angled triangle, with the hypoteneuse equal to the speed c, the shortest side the aether flow speed v, and the other side the cross-stream speed we need to find the time to get across. From Pythagoras' theorem, then, the cross-stream speed is the square root of $(c^2 - v^2)$.

Since this will be the same both ways, the roundtrip cross-stream time will be

$$2w/\sqrt{c^2-v^2}\,.$$

This can be written in the form

$$\frac{2w}{c}\frac{1}{\sqrt{1-v^2/c^2}} \cong \frac{2w}{c}\frac{1}{1-(v^2/2c^2)} \cong \frac{2w}{c}\left(1+\frac{v^2}{2c^2}\right)$$

where the two successive approximations, valid for $v/c = x \ll 1$, are $\sqrt{1-x} \cong 1-(x/2)$ and $1/(1-x) \cong 1 + x$.

Therefore the,

$$\text{cross-stream roundtrip time} \cong \frac{2w}{c} \times \left(1 + \frac{v^2}{2c^2}\right).$$

Looking at the two roundtrip times at the ends of the two paragraphs above, we see that they differ by an amount $(2w/c) \times v^2/2c^2$. Now, $2w/c$ is just the time the light would take if there were no aether wind at all, say, a few millionths of a second. If we take the aether windspeed to be equal to the earth's speed in orbit, for example, v/c is about 1/10,000, so v^2/c^2 is about 1/ 100,000,000.

This means the time delay between the pulses reflected from the different mirrors reaching the telescope is about one-hundred-millionth of a few millionths of a second. It seems completely hopeless that such a short time delay could be detected. However, this turns out *not* to be the case, and Michelson was the first to figure out how to do it.

The trick is to use the interference properties of the lightwaves. Instead of sending pulses of light, as we discussed above, Michelson sent in a steady beam of light of a single colour. This can be visualized as a sequence of ingoing waves, with a wavelength one fifty-thousandth of an inch or so. Now this sequence of waves is split into two, and reflected as previously described. One set of waves goes upstream and downstream, the other goes across stream and back. Finally, they come together into the telescope and the eye.

If the one that took longer is half a wavelength behind, its troughs will be on top of the crests of the first wave, they will cancel, and nothing will be seen. If the delay is less than that, there will still be some dimming. However, slight errors in the placement of the mirrors would have the same effect. This is one reason why the apparatus is built to be rotated. On turning it through 90 degrees, the upstream-downstream and the cross-stream waves change places.

Now the other one should be behind. Thus, if there is an aether wind, if you watch through the telescope while you rotate the turntable, you should expect to see variations in the brightness of the incoming light.

To magnify the time difference between the two paths, in the actual experiment the light was reflected backwards and forwards several times, like a several lap race.

Michelson calculated that an aether windspeed of only one or two miles a second would have observable effects in this experiment, so if the aether windspeed was comparable to the earth's speed in orbit around the sun, it would be easy to see. In fact, *nothing* was observed. The light intensity did not vary at all. Some time later, the experiment was redesigned so that an aether wind caused by the earth's daily rotation could be detected. Again, nothing was seen.

Finally, Michelson wondered if the aether was somehow getting stuck to the earth, like the air in a below-decks cabin on a ship, so he redid the experiment on top of a high mountain in California. Again, no aether wind was observed. It was difficult to believe that the aether in the immediate vicinity of the earth was stuck to it and moving with it, because light rays from stars would deflect as they went from the moving faraway aether to the local stuck aether.

The only possible conclusion from this series of very difficult experiments was that the whole concept of an all-pervading aether was wrong from the start.

Michelson was very reluctant to think along these lines. In fact, new theoretical insight into the nature of light had arisen in the 1860's from the brilliant theoretical work of Maxwell, who had written down a set of equations describing how electric and magnetic fields can give rise to each other. He had discovered that his equations predicted there could be waves made up of electric and magnetic fields, and the speed of these waves, deduced from experiments on how these fields link together, would be 186,300 miles per second.

This is, of course, the speed of light, so it is natural to assume that light is made up of fast-varying electric and magnetic fields. But this leads to a big problem: Maxwell's equations predict a definite speed for light, and it *is* the speed found by measurements. But what is the speed to be measured relative to? The whole point of bringing in the aether was to give a picture for light resembling the one we understand for sound, compressional waves in a medium. The speed of sound through air is measured relative to air. If the wind is blowing towards you from the source of sound, you will hear the sound sooner. If there isn't an aether, though, this analogy doesn't hold up. So what does light travel at 186,300 miles per second relative to? There is another obvious possibility, which is called the emitter theory: the light travels at 186,300 miles per second relative to the source of the light. The analogy here is between light emitted by a source and bullets emitted by a machine gun. The bullets come out at a definite speed (called the muzzle velocity) relative to the barrel of the gun.

If the gun is mounted on the front of a tank, which is moving forward, and the gun is pointing forward, then relative to the ground the bullets are moving faster than they would if shot from a tank at rest. The simplest way to test the emitter theory of light, then, is to measure the speed of light emitted in the forward direction by a flashlight moving in the forward direction, and see if it exceeds the known speed of light by an amount equal to the speed of the flashlight.

Actually, this kind of direct test of the emitter theory only became experimentally feasible in the nineteen-sixties. It is now possible to produce particles, called neutral pions, which decay each one in a little explosion, emitting a flash of light. It is also possible to have these pions moving forward at 185,000 miles per second when they self destruct, and to catch the light emitted in the forward direction, and clock its speed.

It is found that, despite the expected boost from being emitted by a very fast source, the light from the little explosions is going forward at the usual speed of 186,300 miles per second. In the last century, the emitter theory was rejected because it was thought the appearance of certain astronomical phenomena, such as double stars, where two stars rotate around each other, would be affected. Those arguments have since been criticized, but the pion test is unambiguous.

EINSTEIN'S ANSWER

The results of the various experiments discussed above seem to leave us really stuck. Apparently light is not like sound, with a definite speed relative to some underlying medium. However, it is also not like bullets, with a definite speed relative to the source of the light. Yet when we measure its speed we always get the same result. How can all these facts be interpreted in a simple consistent way?

SPECIAL RELATIVITY

GALILEAN RELATIVITY AGAIN

At this point in the course, we finally enter the twentieth century—Albert Einstein wrote his first paper on relativity in 1905. To put his work in context, let us first review just what is meant by "relativity" in physics. The first example, is what is called "Galilean relativity" and is nothing but Galileo's perception that by observing the motion of objects, alive or dead, in a closed room there is no way to tell if the room is at rest or is in fact in a boat moving at a steady speed in a fixed direction.

(You *can* tell if the room is accelerating or turning around.) Everything looks the same in a room in steady motion as it does in a room at rest. After Newton formulated his Laws of Motion, describing how bodies move in response to forces and so on, physicists reformulated Galileo's observation in a slightly more technical, but equivalent, way: they said the laws of physics are the same in a uniformly moving room as they are in a room at rest. In other words, the same force produces the same acceleration, and an object experiencing no force moves at a steady speed in a straight line in either case.

Of course, talking in these terms implies that we have clocks and rulers available so that we can actually time the motion of a body over a measured distance, so the physicist envisions the room in question to have calibrations along all the walls, so the position of anything can be measured, and a good clock to time motion. Such a suitably equipped room is called a "*frame of reference*"—the calibrations on the walls are seen as a frame which you can use to specify the precise position of an object at a given time. (This is the same as a set of "coordinates".) Anyway, the bottom line is that no amount of measuring of motions of objects in the "frame of reference" will tell you whether this is a frame at rest or one moving at a steady velocity.

What exactly do we mean by a frame "at rest" anyway? This seems obvious from our perspective as creatures who live on the surface of the earth—we mean, of course, at rest relative to fixed objects on the earth's surface. Actually, the earth's rotation means this isn't quite a fixed frame, and also the earth is moving in orbit at 18 miles per second. From an astronaut's point of view, then, a frame fixed relative to the sun might seem more reasonable. But why stop there?

We believe the laws of physics are good throughout the universe. Let us consider somewhere in space far from the sun, even far from our galaxy. We would see galaxies in all directions, all moving in different ways. Suppose we now set up a frame of reference and check that Newton's laws still work. In particular, we check that the First Law holds—that a body experiencing no force moves at a steady speed in a straight line. This First law is often referred to as The Principle of Inertia, and a frame in which it holds is called an Inertial Frame.

Then we set up another frame of reference, moving at a steady velocity relative to the first one, and find that Newton's laws are o.k. in this frame too. The point to notice here is that it is not at all obvious which—if either—of these frames is "at rest". We *can*, however, assert that they are both *inertial* frames, after we've checked that in both of them, a body with no forces acting on it moves at a steady speed in a straight line (the speed could be zero). In this situation, Michelson would have said that a frame "at rest" is one at rest relative to the aether. However, his own experiment found motion through the aether to be undetectable, so how would we ever know we were in the right frame?

In the middle of the nineteenth century there was a substantial advance in the understanding of electric and magnetic fields. (In fact, this advance is in large part responsible for the improvement in living standards since that time.)

The new understanding was summarized in a set of equations called Maxwell's equations describing how electric and magnetic fields interact and give rise to each other, just as, two centuries earlier, the new understanding of dynamics was summarized in the set of equations called Newton's laws. The important thing about Maxwell's equations for our present purposes is that they predicted waves made up of electric and magnetic fields that moved at 3×10^8 meters per second, and it was immediately realized that this was no coincidence—light waves must be nothing but waving electric and magnetic fields. (This is now fully established to be the case.)

It is worth emphasizing that Maxwell's work predicted the speed of light from the results of experiments that were not thought at the time they were done to have anything to do with light—experiments on, for example, the strength of electric field produced by waving a magnet. Maxwell was able to deduce a speed for waves like this using methods analogous to those by which earlier scientists had figured out the speed of sound from a knowledge of the density and the springiness of air.

GENERALIZING GALILEAN RELATIVITY TO INCLUDE LIGHT: SPECIAL RELATIVITY

We now come to Einstein's major insight: the Theory of Special Relativity. It is deceptively simple. Einstein first dusted off Galileo's discussion of experiments below decks on a uniformly moving ship, and restated it as:

The Laws of Physics are the same in all Inertial Frames.

Einstein then simply brought this up to date, by pointing out that the Laws of Physics must now include Maxwell's equations describing electric and magnetic fields as well as Newton's laws describing motion of masses under gravity and other forces.

Note for experts and the curious: we shall find that Maxwell's equations are completely unaltered by special relativity, but, as will become clear later, Newton's Laws do need a bit of readjustment to include special relativistic phenomena. The First Law is still o.k., the Second Law in the form $F = ma$ is not, because we shall find mass varies; we need to equate force to rate of change of momentum (Newton understood that, of course—that's the way he stated the law!).

The Third Law, stated as action equals reaction, no longer holds because if a body moves, its electric field, say, does not readjust instantaneously—a ripple travels outwards at the speed of light. Before the ripple reaches another charged body, the electric forces between the two will be unbalanced. However, the crucial consequence of the Third Law—the conservation of momentum when two bodies interact, still holds. It turns out that the rippling field itself carries momentum, and everything balances.) Demanding that Maxwell's equations be satisfied in all inertial frames has one major consequence as far as we are concerned. Maxwell's equations give the speed of light to be 3×10^8 meters per second. Therefore, demanding that the laws of physics are the same in all inertial frames implies that the speed of any light wave, measured in any inertial frame, must be 3×10^8 meters per second.

This then is the entire content of the Theory of Special Relativity: the Laws of Physics are the same in any inertial frame, and, in particular, any measurement of the speed of light in any inertial frame will always give 3×10^8 meters per second.

YOU *REALLY* CAN'T TELL YOU'RE MOVING!

Just as Galileo had asserted that observing gnats, fish and dripping bottles, throwing things and generally jumping around would not help you to find out if you were in a room at rest or moving at a steady velocity, Einstein added that no kind of observation at all, even measuring the speed of light across your room to any accuracy you like, would help find out if your room was "really at rest".

This implies, of course, that the concept of being "at rest" is meaningless. If Einstein is right, there is no natural rest-frame in the universe. Naturally, there can be no "aether", no thin transparent jelly filling space and vibrating with light waves, because if there were, *it* would provide the natural rest frame, and affect the speed of light as measured in other moving inertial frames as discussed above.

So we see the Michelson-Morley experiment was doomed from the start. There never was an aether wind. The light was not slowed down by going "upstream"—light *always* travels at the same speed, which we shall now call c,

$$c = 3\times10^8 \text{ meters per second}$$

to save writing it out every time. This now answers the question of what the speed of light, c, is relative to. We already found that it is not like sound, relative to some underlying medium. It is also not like bullets, relative to the source of the light (the discredited emitter theory). Light travels at c relative to the observer, since if the observer sets up an inertial frame (clocks, rulers, etc.) to measure the speed of light he will find it to be *c*. (We always assume our observers are very competent experimentalists!)

TRUTH AND CONSEQUENCES

The Truth we are referring to here is the seemingly innocuous and plausible sounding statement that all inertial frames are as good as each other—the laws of physics are the same in all of them—and so the speed of light is the same in all of them. This Special Theory of Relativity has some surprising consequences, which reveal themselves most dramatically when things are moving at relative speeds comparable to the speed of light.

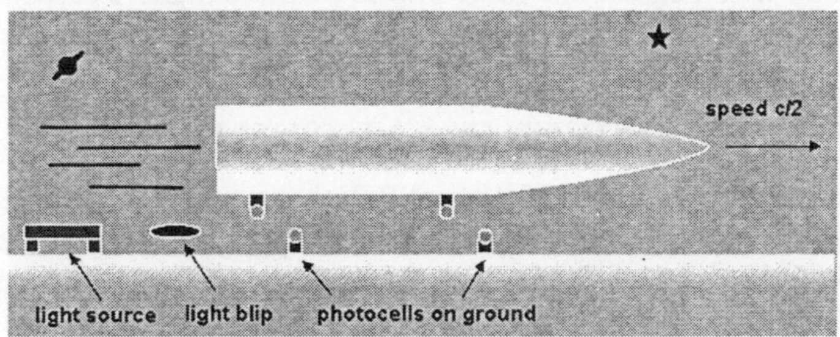

Fig. The spped of the same blip of light is measured by two observers, having relative speed c/2. Both measure the time the the blip takes from one photocell to a second one 10 meters further on. Both find the speed to be c.

Einstein liked to explain his theory using what he called "thought experiments" involving trains and other kinds of transportation moving at these speeds (technically unachievable so far!), and we shall follow his general approach. To begin with, let us consider a simple measurement of the speed of light carried out at the same time in two inertial frames moving at half the speed of light relative to each other. The setup is as follows: on a flat piece of ground, we have a flashlight which emits a blip of light, like a strobe.

We have two photocells, devices which click and send a message down a wire when light falls on them. The photocells are placed 10 meters apart in the path of the blip of light, they are somehow wired into a clock so that the time taken by the blip of light to travel from the first photocell to the second, in other words, the time between clicks, can be measured. From this time and the known distance between them, we can easily find the speed of the blip of light.

Meanwhile, there is another observer, passing overhead in a spaceship traveling at half the speed of light. She is also equipped with a couple of photocells, placed 10 meters apart on the bottom of her spaceship as shown, and she is able to measure the speed of the same blip of light, relative to her frame of reference (the spaceship).

The observer on the spaceship will measure the blip of light to be traveling at c relative to the spaceship, the observer on the ground will measure the same blip to be traveling at c relative to the ground. That is the unavoidable consequence of the Theory of Relativity.

SPECIAL RELATIVITY IN A NUTSHELL

Einstein's Theory of Special Relativity, may be summarized as follows: The Laws of Physics are the same in any Inertial Frame of Reference. (Such frames move at steady velocities with respect to each other.) These Laws include in particular Maxwell's Equations describing electric and magnetic fields, which predict that light always travels at a particular speed c, equal to about 3×10^8 meters per second, that is,186,300 miles per second. It follows that any measurement of the speed of any flash of light by any observer in any inertial frame will give the same answer c.

We have already noted one counter-intuitive consequence of this, that two different observers moving relative to each other, each measuring the speed of the *same* blob of light relative to himself, will *both* get *c*, even if their relative motion is in the same direction as the motion of the blob of light.

We shall now explore how this simple assumption changes everything we thought we understood about time and space.

A SIMPLE BUT RELIABLE CLOCK

We mentioned earlier that each of our (inertial) frames of

reference is calibrated (had marks at regular intervals along the walls) to measure distances, and has a clock to measure time. Let us now get more specific about the clock—we want one that is easy to understand in any frame of reference. Instead of a pendulum swinging back and forth, which wouldn't work away from the earth's surface anyway, we have a blip of light bouncing back and forth between two mirrors facing each other. We call this device a *light clock*.

To really use it as a timing device we need some way to count the bounces, so we position a photocell at the upper mirror, so that it catches the edge of the blip of light. The photocell clicks when the light hits it, and this regular series of clicks drives the clock hand around, just as for an ordinary clock. Of course, driving the photocell will eventually use up the blip of light, so we also need some provision to reinforce the blip occasionally, such as a strobe light set to flash just as it passes and thus add to the intensity of the light. Admittedly, this may not be an easy way to build a clock, but the basic idea is simple.

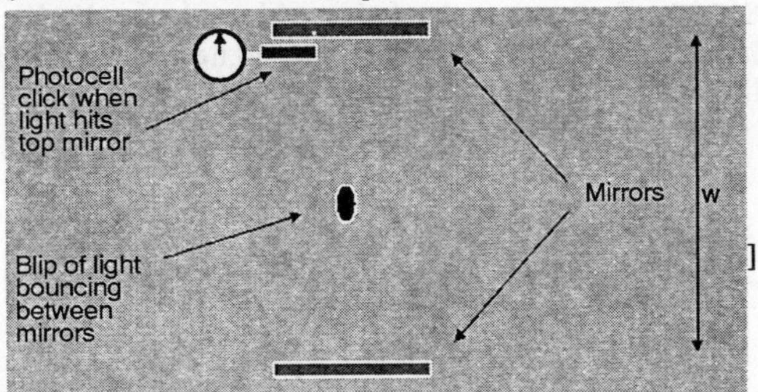

Fig. Einstein's Light Clock: A blip of light bounces between two parallel mirrors w meters apart; each time it hits the top mirror, the photocell moves the clock on one click. Time between clikc $2w/c$.

It's easy to figure out how frequently our light clock clicks. If the two mirrors are a distance w apart, the round trip distance for the blip from the photocell mirror to the other mirror and back is $2w$. Since we know the blip always travels at c, we find the round trip time to be $2w/c$, so this is the time between clicks. This isn't a very long time for a reasonable sized clock! The crystal in a quartz watch "clicks " of the order of 10,000 times a second.

That would correspond to mirrors about nine miles apart, so we need our clock to click about 1,000 times faster than that to get to a reasonable size. Anyway, let us assume that such purely technical problems have been solved.

LOOKING AT SOMEBODY ELSE'S CLOCK

Let us now consider two observers, Jack and Jill, each equipped with a calibrated inertial frame of reference, and a light clock. To be specific, imagine Jack standing on the ground with his light clock next to a straight railroad line, while Jill and her clock are on a large flatbed railroad wagon which is moving down the track at a constant speed v. Jack now decides to check Jill's light clock against his own.

He knows the time for his clock is $2w/c$ between clicks. Imagine it to be a slightly misty day, so with binoculars he can actually see the blip of light bouncing between the mirrors of Jill's clock. How long does he think that blip takes to make a round trip? The one thing he's sure of is that it must be moving at $c = 186{,}300$ miles per second, relative to him—that's what Einstein tells him. So to find the round trip time, all he needs is the round trip distance.

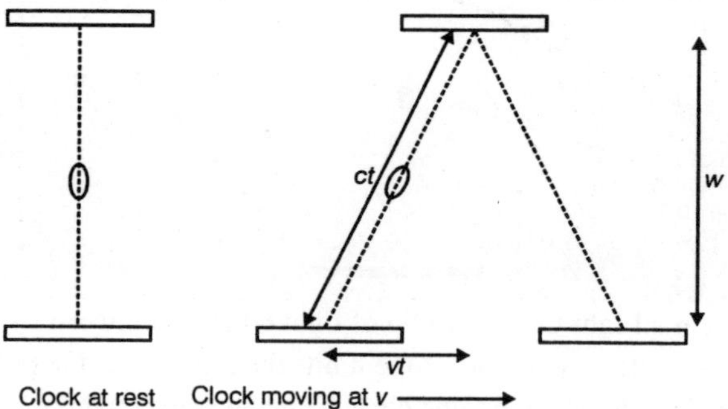

Fig. Two identical light clocks: One at rest, one moving relative to us. The light blips in both travel at the same speed relative to us, the one in the moving clock goes further, so must take longer between clicks. t is the time from one mirror to the other.

This will *not* be $2w$, because the mirrors are on the flatbed wagon moving down the track, so, relative to Jack on the ground, when the blip gets back to the top mirror, that mirror has moved down the track some since the blip left, so the blip actually follows a zigzag path as seen from the ground.

Suppose now the blip in Jill's clock on the moving flatbed wagon takes time t to get from the bottom mirror to the top mirror as measured by Jack standing by the track. Then the length of the "zig" from the bottom mirror to the top mirror is necessarily ct, since that is the distance covered by any blip of light in time t.

Meanwhile, the wagon has moved down the track a distance vt, where v is the speed of the wagon. This should begin to look familiar—it is precisely the same as the problem of the swimmer who swims at speed c relative to the water crossing a river flowing at v! We have again a right-angled triangle with hypotenuse ct, and shorter sides vt and w.

From Pythagoras, then,

$$c^2t^2 = v^2t^2 + w^2$$

so

$$t^2(c^2 - v^2) = w^2$$

or

$$t^2(1 - v^2/c^2) = w^2/c^2$$

and, taking the square root of each side, then doubling to get the round trip time, we conclude that Jack sees the time between clicks for Jill's clock to be:

$$\text{time between clicks for moving clock} = \frac{2w}{c}\frac{1}{\sqrt{1 - v^2/c^2}}$$

Of course, this gives the right answer $2w/c$ for a clock at rest, that is, $v = 0$.

This means that Jack sees Jill's light clock to be going slow—a longer time between clicks—compared to his own identical clock. Obviously, the effect is not dramatic at real railroad speeds. The correction factor is $\sqrt{1 - v^2/c^2}$, which differs from 1 by about one part in a trillion even for a bullet train! Nevertheless, the effect is real and can be measured.

It is important to realise that the only reason we chose a *light* clock, as opposed to some other kind of clock, is that its motion is very easy to analyse from a different frame. Jill could have a collection of clocks on the wagon, and would synchronize them all. For example, she could hang her wristwatch right next to the face of the light clock, and observe them together to be sure they always showed the same time. Remember, in her frame her light clock clicks every $2w/c$ seconds, as it is designed to do. Observing this scene from his position beside the track, Jack will see the synchronized light clock and wristwatch next to each other, and, of course, note that the wristwatch is *also* running slow by the factor $\sqrt{1-v^2/c^2}$ In fact, *all* her clocks, including her pulse, are slowed down by this factor according to Jack. Jill is aging more slowly because she's moving!

But this isn't the whole story—we must now turn everything around and look at it from Jill's point of view. Her inertial frame of reference is just as good as Jack's. She sees his light clock to be moving at speed v (backwards) so from her point of view his light blip takes the longer zigzag path, which means his clock runs slower than hers.

That is to say, each of them will see the other to have slower clocks, and be aging more slowly. This phenomenon is called time dilation. It has been verified in recent years by flying very accurate clocks around the world on jetliners and finding they register less time, by the predicted amount, than identical clocks left on the ground. Time dilation is also very easy to observe in elementary particle physics.

FITZGERALD CONTRACTION

Consider now the following puzzle: suppose Jill's clock is equipped with a device that stamps a notch on the track once a second. How far apart are the notches? From Jill's point of view, this is pretty easy to answer. She sees the track passing under the wagon at v meters per second, so the notches will of course be v meters apart. But Jack sees things differently.

He sees Jill's clocks to be running slow, so he will see the notches to be stamped on the track at intervals of $1/\sqrt{1-v^2/c^2}$ seconds (so for a relativistic train going at $v = 0.8c$, the notches are stamped at intervals of 5/3 = 1.67 seconds). Since Jack agrees with Jill that the relative speed of the wagon and the track is v, he will assert the notches are not v meters apart, but $v/\sqrt{1-v^2/c^2}$ meters apart, a greater distance. Who is right? It turns out that Jack is right, because the notches are in his frame of reference, so he can wander over to them with a tape measure or whatever, and check the distance. This implies that as a result of her motion, Jill observes the notches to be closer together by a factor $\sqrt{1-v^2/c^2}$ than they would be at rest. This is called the Fitzgerald contraction, and applies not just to the notches, but also to the track and to Jack—everything looks somewhat squashed in the direction of motion!

EXPERIMENTAL EVIDENCE FOR TIME DILATION: DYING MUONS

The first clear example of time dilation was provided over fifty years ago by an experiment detecting *muons*. These particles are produced at the outer edge of our atmosphere by incoming cosmic rays hitting the first traces of air.

They are unstable particles, with a "half-life" of 1.5 microseconds (1.5 millionths of a second), which means that if at a given time you have 100 of them, 1.5 microseconds later you will have about 50, 1.5 microseconds after that 25, and so on. Anyway, they are constantly being produced many miles up, and there is a constant rain of them towards the surface of the earth, moving at very close to the speed of light. In 1941, a detector placed near the top of Mount

Washington (at 6000 feet above sea level) measured about 570 muons per hour coming in. Now these muons are raining down from above, but dying as they fall, so if we move the detector to a lower altitude we expect it to detect fewer muons because a fraction of those that came down past the 6000 foot level will die before they get to a lower altitude detector. Approximating their speed by that of light, they are raining down at 186,300 miles per second, which turns out to be, conveniently, about 1,000 feet per microsecond.

Thus they should reach the 4500 foot level 1.5 microseconds after passing the 6000 foot level, so, if half of them die off in 1.5

microseconds, as claimed above, we should only expect to register about 570/2 = 285 per hour with the same detector at this level. Dropping another 1500 feet, to the 3000 foot level, we expect about 280/2 = 140 per hour, at 1500 feet about 70 per hour, and at ground level about 35 per hour.

To summarize: given the known rate at which these raining-down unstable muons decay, and given that 570 per hour hit a detector near the top of Mount Washington, we only expect about 35 per hour to survive down to sea level. In fact, when the detector was brought down to sea level, it detected about 400 per hour! How did they survive?

The reason they didn't decay is that in their frame of reference, much less time had passed. Their actual speed is about 0.994c, corresponding to a time dilation factor of about 9, so in the 6 microsecond trip from the top of Mount Washington to sea level, their clocks register only 6/9 = 0.67 microseconds. In this period of time, only about one-quarter of them decay.

What does this look like from the muon's point of view? How do they manage to get so far in so little time? To them, Mount Washington and the earth's surface are approaching at 0.994c, or about 1,000 feet per microsecond. But in the 0.67 microseconds it takes them to get to sea level, it would seem that to them sea level could only get 670 feet closer, so how could they travel the whole 6000 feet from the top of Mount Washington? The answer is the Fitzgerald contraction.

To them, Mount Washington is squashed in a vertical direction (the direction of motion) by a factor of $\sqrt{1-v^2/c^2}$ the same as the time dilation factor, which for the muons is about 9. So, to the muons, Mount Washington is only 670 feet high—this is why they can get down it so fast.

SPECIAL RELATIVITY: SYNCHRONIZING CLOCKS

Suppose we want to synchronize two clocks that are some distance apart.

We could stand beside one of them and look at the other through a telescope, but we'd have to remember in that case that we are seeing the clock as it was when the light left it, and correct accordingly. Another way to be sure the clocks are synchronized, assuming they are both accurate, is to start them together. How can we do that? We could, for example, attach a photocell to each clock, so when a flash of light reaches the clock, it begins running.

The clocks are triggered when the flash of light from the central bulb reaches the attached photocells.

Fig. The clocks are Triggered when the flash of light from the central bulb reaches the attached photocells.

If, then, we place a flashbulb at the midpoint of the line joining the two clocks, and flash it, the light flash will take the same time to reach the two clocks, so they will start at the same time, and therefore be synchronized.

Let us now put this whole arrangement - the two clocks and the midpoint flashbulb - on a train, and we suppose the train is moving at some speed v to the right, say half the speed of light or so. Let's look carefully at the clock-synchronizing operation as seen from the ground.

In fact, an observer on the ground would say the clocks are not synchronized by this operation! The basic reason is that he would see the flash of light from the middle of the train traveling at c relative to the ground in each direction, but he would also observe the back of the train coming at v to meet the flash, whereas the front is moving at v away from the bulb, so the light flash must go further to catch up.

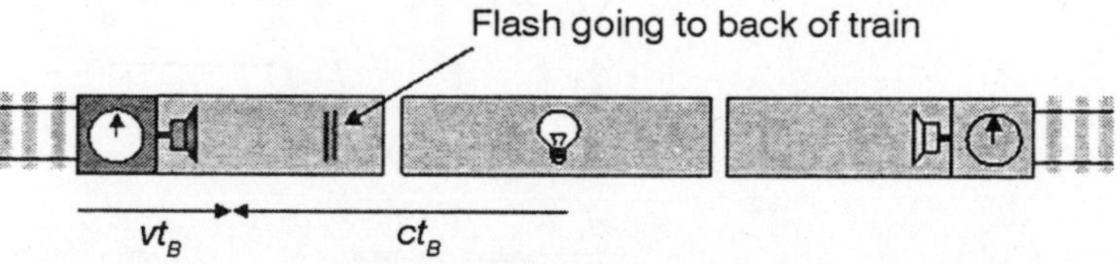

Fig. The train is moving to the right: the central bulb emits a flash of light. Seen from the ground, the part of the flash moving towards the rear travels at c, the rear travels at v to meet it.

In fact, it is not difficult to figure out how much later the flash reaches the front of the train compared with the back of the train, as viewed from the ground. First recall that as viewed from the ground the train has length $L\sqrt{1-v^2/c^2}$.

The train is moving to the right: the central bulb emits a flash of light. Seen from the ground, the part of the flash moving towards the rear travels at c, the rear travels at v to meet it.

Letting t_B be the time it takes the flash to reach the back of the train, it is clear from the figure that

$$vt_B + ct_B = \frac{L}{2}\sqrt{1-\frac{v^2}{c^2}}$$

from which t_B is given by

$$t_B = \frac{1}{c+v}\frac{L}{2}\sqrt{1-\frac{v^2}{c^2}}$$

In a similar way, the time for the flash of light to reach the front of the train is (as measured by a ground observer)

$$t_F = \frac{1}{c-v}\frac{L}{2}\sqrt{1-\frac{v^2}{c^2}}$$

Therefore the time difference between the starting of the two clocks, as seen from the ground, is

$$t_F - t_B = \left(\frac{1}{c-v} - \frac{1}{c+v}\right)\frac{L}{2}\sqrt{1-(v/c)^2}$$

$$= \left(\frac{2v}{c^2-v^2}\right)\frac{L}{2}\sqrt{1-(v/c)^2}$$

$$= \frac{2v}{c^2}\frac{1}{1-(v-c)^2}\frac{L}{2}\sqrt{1-(v/c)^2}$$

$$= \frac{vL}{c^2}\frac{1}{\sqrt{1-(v/c)^2}}$$

Remember, this is the time difference between the starting of the train's back clock and its front clock as measured by an observer on the ground with clocks on the ground. However, to this observer the clocks on the train appear to tick more slowly, by the factor $\sqrt{1-(v/c)^2}$, so that although the ground observer measures the time interval between the starting of the clock at the back of the train and the clock at the front as $(vL/c^2)\left(1/\sqrt{1-(v/c)^2}\right)$ seconds, he also sees the slow running clock at the back actually reading vL/c^2 seconds at the instant he sees the front clock to start.

To summarize: as seen from the ground, the two clocks on the train (which is moving at v in the x-direction) are running slowly, registering only $\sqrt{1-(v/c)^2}$ seconds for each second that passes. Equally important, the clocks—which are synchronized by an observer on the train—appear unsynchronized when viewed from the ground, the one at the back of the train reading vL/c^2 seconds ahead of the clock at the front of the train, where L is the rest length of the train (the length as measured by an observer on the train).

Note that if $L = 0$, that is, if the clocks are together, both the observers on the train and those on the ground will agree that they are synchronized. We need a *distance* between the clocks, as well as relative motion, to get a disagreement about synchronization.

THE LORENTZ TRANSFORMATIONS

PROBLEMS WITH THE GALILEAN TRANSFORMATIONS

We have already seen that Newtonian mechanics is invariant under the *Galilean* transformations relating two inertial frames moving with relative speed v in the x-direction,

$$x = x' + vt'$$
$$y = y'$$
$$z = z'$$
$$t = t'.$$

However, these transformations presuppose that *time* is a well-defined universal concept, that is to say, it's the same time everywhere, and all observers can agree on what time it is. Once we accept the basic postulate of special relativity, however, that the laws of physics, including Maxwell's equations, are the same in all inertial frames of reference, and consequently the speed of light has the same value in all inertial frames, then as we have seen, observers in different frames do *not* agree on whether clocks some distance apart are synchronized.

Furthermore, as we have discussed, measurements of moving objects are compressed in the direction of motion by the Lorentz-Fitzgerald contraction effect. Obviously, the above equations are too naove! We must think more carefully about time and distance measurement, and construct new transformation equations consistent with special relativity.

Our aim here, then, is to find a set of equations analogous to those above giving the coordinates of an *event* (x, y, z, t) in frame S, for example, a small bomb explosion, as functions of the coordinates (x,y, z, t) of the same event measured in the parallel frame S' which is moving at speed v along the x-axis of frame S. Observers O at the origin in frame S and O' at the origin in frame S synchronize their clocks at $t = t' = 0$, at the instant they pass each other, that is, when the two frames coincide. (Using our previous notation, O is Jack and O' is Jill.)

To determine the time t' at which the bomb exploded in her frame, O' could determine the distance of the point (x', y', z') from the origin, and hence how long it would take light from the explosion to reach her at the origin. A more direct approach (which is helpful in considering transformations between different frames) is to imagine O' to have a multitude of helpers, with an array of clocks throughout the frame, which have all been synchronized by midpoint flashes. Then the event—the bomb explosion—will be close to a clock, and that local clock determines the time t' of the event, so we do not need to worry about timing a light signal.

In frame S', then, O' and her crew have clocks all along the x'-axis (as well as everywhere else) and all synchronized:

Now consider how this string of clocks appears as viewed by O from frame S. First, since they are all moving at speed v, they will be registering time more slowly by the usual time dilation factor $\sqrt{1-v^2/c^2}$ than O's own physically identical clocks. Second, they will not be synchronized. The clocks are L apart as measured by O', successive clocks to the right (the direction of motion) will be behind by Lv/c^2 as observed by O.

It should be mentioned that this lack of synchronization as viewed from another frame only occurs for clocks separated in the direction of relative motion. Consider two clocks some distance apart on the z' axis of S'. If they are synchronized in S' by both being started by a flash of light from a bulb half way between them, it is clear that as viewed from S the light has to go the same distance to each of the clocks, so they will still be synchronized (although they will start later by the time dilation factor).

DERIVING THE LORENTZ TRANSFORMATIONS

Let us now suppose that O' and her crew observe a small bomb to explode in S' at $(x', 0, 0, t')$. The space coordinates and time (x, y, z, t) of this event as observed by O in the frame S. (As above, S' moves relative to S at speed v along the x-axis). In other words, we shall derive the Lorentz transformations—which are just the equations giving the four coordinates of an event in one inertial frame in terms of the coordinates of the same event in another inertial frame. We take y', z' zero because they transform trivially—there is no Lorentz contraction perpendicular to the motion, so $y = y'$ and $z = z'$.

First, we consider at what *time* the bomb explodes as measured by O. O' 's crew found the bomb to explode at time t' as measured by a local clock, that is, one located at the site of the explosion, x'. Now, as observed by O from frame S, O's clock at x' is *not* synchronized with O' 's clock at the origin S'. When the bomb explodes and the clock at x' reads t', O will see O' 's origin clock to read $t' + vx'/c^2$.What does O' s own clock read at this point? Recall that O, O' synchronized their origin clocks at the moment they were together, at $t + t' = 0$. Subsequently, O will have observed O' 's clock to be running slowly by the time-dilation factor. Therefore, when at the instant of the explosion he sees O''s origin clock to be reading $t' + vx'/c^2$, he will find that the true time t in his frame is equal to this appropriately scaled to allow for time dilation, that is,

$$t = \frac{t' + vx'/c^2}{\sqrt{1 - v^2/c^2}}.$$

This is the first of the Lorentz transformations.

The second question is: *where* does O observe the explosion to occur?

Since it occurs at time t after O' passed O, O' is vt meters beyond O at the time of the explosion. The explosion takes place x' meters beyond O', as measured by O', but of course O will see that distance x' as contracted to $x'\sqrt{1 - v^2/c^2}$ since it's in a moving frame.

Therefore O observes the explosion at point x given by

$$x = vt + x'\sqrt{1 - v^2/c^2}.$$

This can be written as an equation for x in terms of x', t' by substituting for t using the first Lorentz transformation above, to give

$$x = \frac{x' + vt'}{\sqrt{1 - v^2/c^2}}$$

Therefore, we have found the Lorentz transformations expressing the coordinates (x, y, z, t) of an event in frame S in terms of the coordinates (x', y', z', t') of the same event in frame S':

$$x = \frac{x' + vt'}{\sqrt{1 - v^2/c^2}}$$

$$y = y'$$

$$z = z'$$

$$t = \frac{t' + vx'/c^2}{\sqrt{1 - v^2/c^2}}$$

Notice that nothing in the above derivation depends on the

x-velocity v of S' relative to S being positive. Therefore, the inverse transformation (from (x,y,z,t) to (x', y', z', t')) has exactly the same form as that given above with v replaced by $-v$.

SPHERES OF LIGHT

Consider now the following scenario: suppose that as O' passes O (the instant both of them agree is at time $t' = t = 0$) O' flashes a bright light, which she observes to create an expanding spherical shell of light, centered on herself (imagine it' s a slightly foggy day, so she can see how the ripple of light travels outwards). At time t', then, O' (or, to be precise, her local observers out there in the frame) will see a shell of light of radius ct', that is to say, they will see the light to have reached all points (x', y', z') on the surface

$$x'^2 + y'^2 + z'^2 = c^2t'^2.$$

Question: how do O and his observers stationed throughout the frame S see this light as rippling outwards?

To answer this question, notice that the above equation for where the light is in frame S' at a particular time t' can be written

$$x'^2 + y'^2 + z'^2 - c^2t'^2 = 0,$$

and can be thought of as a surface in the four dimensional (x', y', z', t') space, the totality of all the "events" of the light reaching any particular point. Now, to find the corresponding surface of events in the four dimensional (x, y, z, t) space, all we have to do is to change from one set of variables to the other using the Lorentz transformations:

$$x' = \frac{x + vt}{\sqrt{1 - v^2/c^2}}$$

$$y' = y$$

$$z' = z$$

$$t' = \frac{t + vx/c^2}{\sqrt{1 - v^2/c^2}}$$

On putting these values of (x', y', z', t') into $x'^2 + y'^2 + z'^2 - c^2t'^2 = 0$ we find that the corresponding surface of events in (x, y, z, t) space is:

$$\text{p}\, x^2 + y^2 + z^2 - c^2t^2 = 0.$$

This means that at time t, O and his observers in frame S will say the light has reached a spherical surface centered on O.

How can O' and O, as they move further apart, possibly both be right in maintaining that at any given instant the outward moving light pulse has a spherical shape, each saying it is centered on herself or himself?

Imagine the light shell as O' sees it—at the instant t' she sees a sphere of radius r', in particular she sees the light to have reached the spots $+r'$ and $-r'$ on the x' axis. But from O' 's point of view the expanding light sphere does *not* reach the point $+r'$ at the same time it reaches $-r'$! (This is just the old story of synchronizing the two clocks at the front and back of the train one more time.) That is why O does not see O' 's sphere: *the* arrival of the light at the sphere of radius r' around O' at time t' corresponds in S to a continuum of different events happening at different times.

LORENTZ INVARIANTS

We found above that for an event (x', y', z', t') for which $x'^2 + y'^2 + z'^2 - c^2t'^2 = 0$ the coordinates of the event (x, y, z, t) as measured in the other frame S satisfy $x^2 + y^2 + z^2 - c^2t^2 = 0$ The quantity $x^2 + y^2 + z^2 - c^2t^2$ is said to be a Lorentz invariant: it doesn't vary on going from one frame to another.

A simple two-dimensional analogy to this invariant is given by considering two sets of axes, Oxy and $Ox'\,y'$ having the same origin O, but the axis Ox' is at an angle to Ox, so one set of axes is the same as the other set but rotated. The point P with coordinates (x,y) has coordinates (x', y') measured on the $Ox'\,y'$ axes.

The square of the distance of the point P from the common origin O is $x^2 + y^2$ and is also $x'^2 + y'^2$, so for the transformation from coordinates (x, y) to (x', y'), $x^2 + y^2$ is an invariant. Similarly, if a point P_1 has coordinates (x_1, y_1) and (x_12, y_12) and another point P_2 has coordinates (x_2, y_2) and (x_22, y_22) then clearly the two points are the same distance apart as measured with respect to the two sets of axes, so

$$(x_1 - x_2)^2 + (y_1 - y_2)^2 = (x_1' - x_2')^2 + (y_1' - y_2')^2.$$

This is really obvious: the distance between two points in an ordinary plane can't depend on the angle at which we choose to set our coordinate axes.

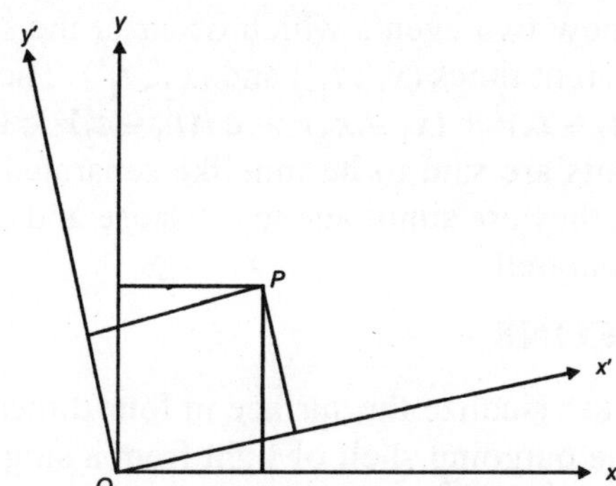

The Lorentz analog of this, dropping the *y,z* coordinates, can be written

$$c^2(t_1 - t_2)^2 + (x_1 - x_2)^2 = c^2(t'_1 - t'_2)^2 + (x'_1 - x'_2)^2 = s^2.$$

say, where *s* is some sort of measure of the "distance" between the two events (x_1, t_1) and (x_2, t_2).

This s is sometimes called the "space-time interval". The big difference from the two-dimensional rotation case is that s^2 *can be positive or negative*. For s^2 negative just taking *s* as its square root the "distance" itself would be imaginary—so we need to think this through carefully. It's clear enough what to do in any particular case—the cases of spacelike and timelike separated events are best dealt with separately, at least to begin with.

Consider first two events simultaneous in frame S', so $t_1' = t_2'$. They will not be simultaneous in frame *S*, but they *will* satisfy

$$(x_1 - x_2)^2 - c^2(t_1 - t_2)^2 > 0.$$

We say the two events are *spacelike separated*. This means that they are sufficiently removed spatially that a light signal could not have time between them to get from one to the other, so one of these events could not be the *cause* of the other. The sequence of two events can be different in different frames if the events are spacelike separated.

Consider again the starting of the two clocks at the front and back of a train as seen from the ground: the back clock starts first. Now imagine viewing this from a faster train overtaking the clock train—now the front clock will be the first to start. The important point is that although these events appear to occur in a different order in a different frame, neither of them could be the cause of the other, so cause and effect are not switched around.

Consider now two events which occur at the same place in frame S' at different times (x_1', t_1') and (x_2', t_1'). Then in frame S:

$$c^2(t_1 - t_2)^2 + (x_1 - x_2)^2 = c^2(t'_1 - t'_2)^2 > 0.$$

These events are said to be timelike separated. There is no frame in which they are simultaneous. "Cause and effect" events are timelike separated.

THE LIGHT CONE

Let us try to visualize the surface in four-dimensional space described by the outgoing shell of light from a single flash,

$$x^2 + y^2 + z^2 - c^2t^2 = 0.$$

It is helpful to think about a simpler situation, the circular ripple spreading on the surface of calm water from a pebble falling in. Taking c here to be the speed of the water waves, it is easy to see that at time t after the splash the ripple is at

$$x^2 + y^2 - c^2t^2 = 0.$$

Now think about this as a surface in the three-dimensional space (x, y, t). The plane corresponding to time t cuts this surface in a circle of radius ct. This means the surface is a cone with its point at the origin. The four-dimensional space flash-of-light surface is not so easy to visualize, but is clearly the higher-dimensional analog: the plane surface corresponding to time t cuts it in a sphere instead of a circle. This surface is called the *lightcone*.

We have stated above that the separation of a point $P\,(x, y, z, t)$ from the origin is spacelike if $x^2 + y^2 + z^2 - c^2t^2 > 0$.and timelike if $x^2 + y^2 + z^2 - c^2t^2 < 0$. It is said to be lightlike if $x^2 + y^2 + z^2 - c^2t^2 = 0$.Points on the light cone described above are lightlike separated from the origin. To be precise, the points corresponding to an outgoing shell of light from a flash at the origin at $t = 0$ form the forward light cone. Since the equation depends only on t^2, there is a solution with t negative, the "backward light cone", just the reflection of the forward light cone in the plane $t = 0$.

Possible causal connections are as follows: an event at the origin (0, 0, 0, 0) could cause an event in the forward light cone: so that is the "future", as seen from the origin.

Events in the backward light cone—the "past"—could cause an event at the origin. There can be no causal link between an event at the origin and an event outside the light cones, since the separation is spacelike: outside the light cones is "elsewhere" as viewed from the origin.

TIME DILATION: A WORKED EXAMPLE

"Moving Clocks Run Slow" *plus* "Moving Clocks Lose Synchronization" *plus* "Length Contraction" leads to consistency!

The object of this exercise is to show explicitly how it is possible for two observers in inertial frames moving relative to each other at a relativistic speed to each see the other's clocks as running slow and as being unsynchronized, and yet if they both look at the same clock at the same time from the same place (which may be far from the clock), they will agree on what time it shows!

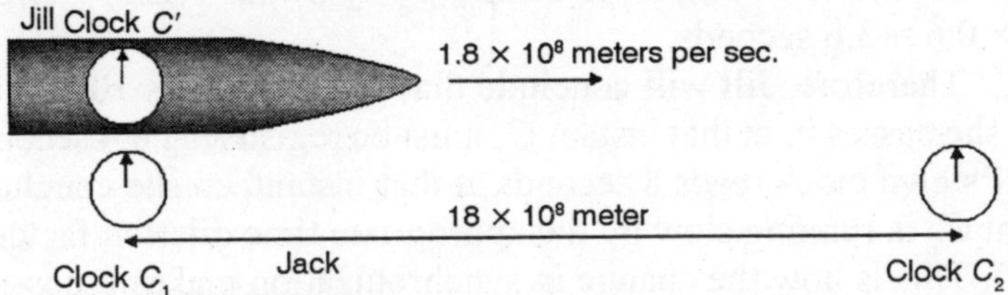

Fig. Jill in her Relativistic rocket passes Jack's first clock at an instant when both their clocks read zero.

Suppose that in Jack's frame we have two synchronized clocks C_1 and C_2 set 18 x 10^8 meters apart (that's about a million miles, or 6 light-seconds). Jill's spaceship, carrying a clock C', is traveling at 0.6*c*, that is 1.8 x 10^8 meters per second, parallel to the line C_1C_2, passing close by each clock.

Suppose C' is synchronized with C_1 as they pass, so both read zero. As measured by Jack the spaceship will take just 10 seconds to reach C_2, since the distance is 6 light seconds, and the ship is traveling at 0.6*c*.

What does clock C' (the clock on the ship) read as it passes C_2?

The time dilation factor

$$\sqrt{1-(v^2/c^2)} = 4/5$$

so C', Jill's clock, will read 8 seconds.

Thus if both Jack and Jill are at C_2 as Jill and her clock C' pass C_2, both will agree that the clocks look like:

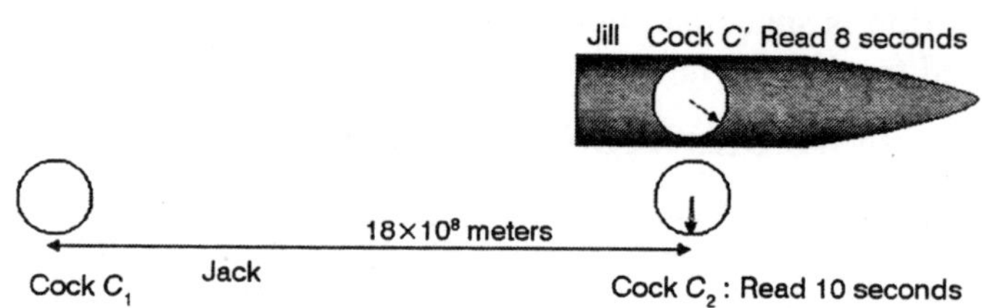

Fig. As Jill passes Jack's second clock, both see that his clock reads 10 seconds, hers reads 8 secons.

How, then, can Jill claim that Jack's clocks C_1, C_2 are the ones that are running slow?

To Jill, C_1, C_2 are running slow, but remember they are *not* synchronized. To Jill, C_1 is behind C_2 by $Lv/c2 = (L/c) \times (v/c) = 6 \times 0.6 = 3.6$ seconds.

Therefore, Jill will conclude that since C_2 reads 10 seconds as she passes it, at that instant C_1 must be registering 6.4 seconds. Jill's own clock reads 8 seconds at that instant, so she concludes that C_1 is running slow by the appropriate time dilation factor of *4/5*. This is how the change in synchronization makes it possible for both Jack and Jill to see the other's clocks as running slow.

Of course, Jill's assertion that as she passes Jack's second "ground" clock C_2 the first "ground" clock C_1 must be registering 6.4 seconds is not completely trivial to check! After all, that clock is now a million miles away! Let us imagine, though, that both observers are equipped with Hubble-style telescopes attached to fast acting cameras, so reading a clock a million miles away is no trick. To settle the argument, the two of them agree that as she passes the second clock, Jack will be stationed at the second clock, and at the instant of her passing they will both take telephoto digital snapshots of the faraway clock C_1, to see what time it reads.

Jack, of course, knows that C_1 is 6 light seconds away, and is synchronized with C_2 which at that instant is reading 10 seconds, so his snapshot must show C_1 to read 4 seconds. That is, looking at C_1 he sees it as it was six seconds ago.

What does Jill's digital snapshot show? It must be identical—two snapshots taken from the same place at the same time must show the same thing! So, Jill must also gets a picture of C_1 reading 4 seconds.

How can she reconcile a picture of the clock reading 4 seconds with her assertion that at the instant she took the photograph the clock was registering 6.4 seconds?

The answer is that she can if she knows her relativity!

First point: length contraction. To Jill, the clock C_1 is actually only 4/5 x 18 x 10^8 meters away (she sees the distance C_1C_2 to be Lorentz contracted!).

Second point: The light didn't even have to go that far! In her frame, the clock C_1 is *moving away*, so the light arriving when she's at C_2 must have left C_1 when it was closer—at distance x in the figure below. The figure shows the light in her frame moving from the clock towards her at speed c, while at the same time the clock itself is moving to the left at $0.6c$.

It might be helpful to imagine yourself in her frame of reference, so you are at rest, and to think of clocks C_1 and C_2 as being at the front end and back end respectively of a train that is going past you at speed $0.6c$. Then, at the moment the back of the train passes you, you take a picture (through your telescope, of course) of the clock at the front of the train. Obviously, the light from the front clock that enters your camera at that instant left the front clock some time ago.

During the time that light traveled towards you at speed c, the front of the train itself was going in the opposite direction at speed $0.6c$. But you know the length of the train in your frame is $4/5 \times 18 \times 10^8$ meters, so since at the instant you take the picture the back of the train is passing you, the front of the train must be $4/5 \times 18 \times 10^8$ meters away. Now that distance, $4/5 \times 18 \times 10^8$, is the sum of the distance the light entering your camera traveled plus the distance the train traveled in the same time, that is, $(1 + 0.6)/1$ times the distance the light traveled.

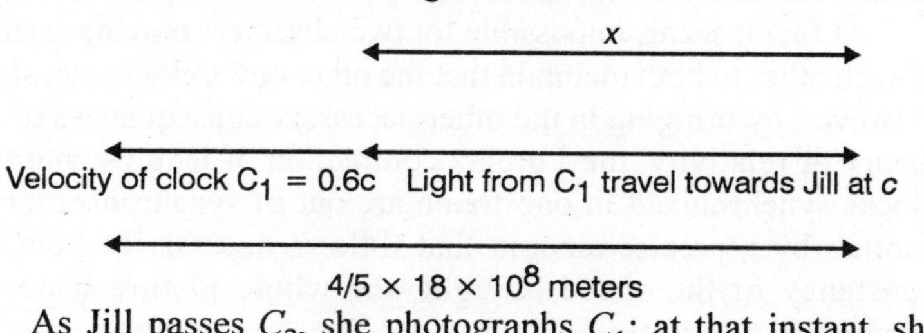

As Jill passes C_2, she photographs C_1: at that instant, she knows C_1 is $4/5 \times 18 \times 10^8$ meters away in her frame, but the light reaching her camera at that moment left C_1 when it was at a distance x, not so far away. As the light traveled towards her at speed c, C_1 was moving away at a speed of 0.6c, so the distance $4/5 \times 10 \times 10^8$ meters is the sum of how far the light traveled towards her and how far the clock traveled away from her, both starting at x.

So the image of the first ground clock she sees and records as she passes the second ground clock must have been emitted when the first clock was a distance x from her in her frame, where

$$x(1 + 3/5) = 4/5 \times 10 \times 10^8 \text{ meters, so } x = 9 \times 10^8 \text{ meters.}$$

Having established that the clock image she is seeing as she takes the photograph left the clock when it was only 9×10^8 meters away, that is, 3 light seconds, she concludes that she is observing the first ground clock as it was three seconds ago.

Third point: time dilation. The story so far: she has a photograph of the first ground clock that shows it to be reading 4 seconds. She knows that the light took three seconds to reach her. So, what can she conclude the clock must actually be registering at the instant the photo was taken? If you are tempted to say 7 seconds, you have forgotten that in her frame, the clock is moving at $0.6c$ and hence *runs slow* by a factor 4/5. Including the time dilation factor correctly, she concludes that in the 3 seconds that the light from the clock took to reach her, the clock itself will have ticked away $3 \times 4/5$ seconds, or 2.4 seconds.

Therefore, since the photograph shows the clock to read 4 seconds, and she finds the clock must have run a further 2.4 seconds, she deduces that at the instant she took the photograph the clock must actually have been registering 6.4 seconds, which is what she had claimed all along!

At first it seems impossible for two observers moving relative to each other to both maintain that the other one's clocks run slow. However, by bringing in the other necessary consequences of the theory of relativity, the Lorentz contraction of lengths, and that clocks synchronized in one frame are out of synchronization in another by a precise amount that follows necessarily from the constancy of the speed of light, the whole picture becomes completely consistent.

MORE RELATIVITY: THE TRAIN AND THE TWINS

EINSTEIN'S DEFINITION OF COMMON SENSE

Although Einstein's theory of special relativity solves the problem posed by the Michelson-Morley experiment—the nonexistence of an ether—it is at a price. The simple assertion that the speed of a flash of light is always c in any inertial frame leads to consequences that defy common sense. When this was pointed out somewhat forcefully to Einstein, his response was that common sense is the layer of prejudices put down before the age of eighteen.

All our intuition about space, time and motion is based on childhood observation of a world in which no objects move at speeds comparable to that of light. Perhaps if we had been raised in a civilization zipping around the universe in spaceships moving at relativistic speeds, Einstein's assertions about space and time would just seem to be common sense. The real question, from a scientific point of view, is not whether special relativity defies common sense, but whether it can be shown to lead to a contradiction. If that is so, common sense wins. Ever since the theory was published, people have been writing papers claiming it *does* lead to contradictions.

The worked example on time dilation, shows how careful analysis of an apparent contradiction leads to the conclusion that in fact there was no contradiction after all. We shall consider other apparent contradictions and think about how to resolve them. This is the best way to build up an understanding of relativity.

TRAPPING A TRAIN IN A TUNNEL

One of the first paradoxes to be aired was based on the Fitzgerald contraction. Recall that any object moving relative to an observer will be seen by that observer to be contracted, foreshortened in the direction of motion by the ubiquitous factor $\sqrt{1-v^2/c^2}$. Einstein lived in Switzerland, a very mountainous country where the railroads between towns often go through tunnels deep in the mountains.

Suppose a train of length L is moving along a straight track at a relativistic speed and enters a tunnel, also of length L. There are bandits inhabiting the mountain above the tunnel. They observe a short train, one of length $L\sqrt{1-v^2/c^2}$, so they wait until this short train is completely inside the tunnel of length L, then they close doors at the two ends, and the train is trapped fully inside the mountain. Now look at this same scenario from the point of view of someone on the train. He sees a train of length L, approaching a tunnel of length $L\sqrt{1-v^2/c^2}$,so the tunnel is not as long as the train from his viewpoint! What does he think happens when the bandits close both the doors?

THE TUNNEL DOORS ARE CLOSED SIMULTANEOUSLY

The key to understanding what is happening here is that we said the bandits closed the two doors at the ends of the tunnel *at the same time*. How could they arrange to do that, since the doors are far apart?

They could use walkie-talkies, which transmit radio waves, or just flash a light down the tunnel, since it's long and straight. Remember, though, that the train is itself going at a speed close to that of light, so they have to be quite precise about this timing! The simplest way to imaging them synchronizing the closings of the two doors is to assume they know the train's timetable, and at a prearranged appropriate time, a light is flashed halfway down the tunnel, and the end doors are closed when the flash of light reaches the ends of the tunnel. Assuming the light was positioned correctly in the middle of the tunnel, that should ensure that the two doors close simultaneously.

OR ARE THEY?

Now consider this door-closing operation from the point of view of someone on the train. Assume he's in an observation car and has incredible eyesight, and there's a little mist, so he actually sees the light flash, and the two flashes traveling down the tunnels towards the two end doors. Of course, the train is a perfectly good inertial frame, so he sees these two flashes to be traveling in opposite directions, but both at c, relative to the train. Meanwhile, he sees the tunnel itself to be moving rapidly relative to the train. Let us say the train enters the mountain through the "front" door. The observer will see the door at the other end of the tunnel, the "back" door, to be rushing towards him, and rushing to meet the flash of light. Meanwhile, once he's in the tunnel, the front door is receding rapidly behind him, so the flash of light making its way to that door has to travel further to catch it. So the two flashes of light going down the tunnel in opposite directions do not reach the two doors simultaneously as seen from the train.

The concept of simultaneity, events happening at the same time, is not invariant as we move from one inertial frame to another. The man on the train sees the back door close first, and, if it is not quickly reopened, the front of the train will pile into it before the front door is closed behind the train.

DOES THE FITZGERALD CONTRACTION WORK SIDEWAYS?

The above discussion is based on Einstein's prediction that objects moving at relativistic speed appear shrunken in their direction of motion. How do we know that they're not shrunken in all three directions, i.e. moving objects maybe keep the same shape, but just get smaller? This can be seen *not* to be the case through a symmetry argument, also due to Einstein. Suppose two trains traveling at equal and opposite relativistic speeds, one north, one south, pass on parallel tracks. Suppose two passengers of equal height, one on each train, are standing leaning slightly out of open windows so that their noses should very lightly touch as they pass each other.

Now, if *N* (the northbound passenger) sees *S* as shrunken in height, *N*'s nose will brush against *S*'s forehead, say, and *N* will feel *S*'s nose brush his chin. Afterwards, then, *N* will have a bruised chin (plus nose), *S* a bruised forehead (plus nose). But this is a perfectly symmetric problem, so *S* would say *N* had the bruised forehead, etc. They can both get off their trains at the next stations and get together to check out bruises.

They must certainly be symmetrical! The only *consistent symmetrical solution* is given by asserting that *neither* sees the other to shrink in height (i.e. in the direction perpendicular to their relative motion), so that their noses touch each other. Therefore, the Lorentz contraction *only* operates in the direction of motion, objects get squashed but not shrunken.

HOW TO GIVE TWINS VERY DIFFERENT BIRTHDAYS

Perhaps the most famous of the paradoxes of special relativity, which was still being hotly debated in national journals in the fifties, is the twin paradox. The scenario is as follows. One of two twins - the sister—is an astronaut. (Flouting tradition, we will take fraternal rather than identical twins, so that we can use "he" and "she" to make clear which twin we mean). She sets off in a relativistic spaceship to alpha-centauri, four light-years away, at a speed of, say, $0.6c$.

When she gets there, she immediately turns around and comes back. As seen by her brother on earth, her clocks ran slowly by the time dilation factor $\sqrt{1-v^2/c^2}$, so although the round trip took 8/0.6 years = 160 months by earth time, she has only aged by 4/5 of that, or 128 months. So as she steps down out of the spaceship, she is 32 months younger than her twin brother.

But wait a minute—how does this look from *her* point of view? She sees the earth to be moving at $0.6c$, first away from her then towards her. So she must see her brother's clock on earth to be running slow! So doesn't she expect her brother on earth to be the younger one after this trip?

The key to this paradox is that this situation is not as symmetrical as it looks. The two twins have quite different experiences. The one on the spaceship is *not* in an inertial frame during the initial acceleration *and* the turnaround and braking periods. (To get an idea of the speeds involved, to get to $0.6c$ at the acceleration of a falling stone would take over six months.) Our analysis of how a clock in one inertial frame looks as viewed from another doesn't work during times when one of the frames isn't inertial - in other words, when one is accelerating.

THE TWINS STAY IN TOUCH

To try to see just how the difference in ages might develop, let us imagine that the twins stay in touch with each other throughout the trip. Each twin flashes a powerful light once a month, according to their calendars and clocks, so that by counting the flashes, each one can monitor how fast the other one is aging.

The questions we must resolve are:

If the brother, on earth, flashes a light once a month, how frequently, as measured by her clock, does the sister see his light to be flashing as she moves away from earth at speed $0.6c$?

How frequently does she see the flashes as she is *returning* at $0.6c$?

How frequently does the brother on earth see the flashes from the spaceship?

Once we have answered these questions, it will be a matter of simple bookkeeping to find how much each twin has aged.

FIGURING THE OBSERVED TIME BETWEEN FLASHES

To figure out how frequently each twin observes the other's flashes to be, on time dilation. In some ways, that was a very small scale version of the present problem. Recall that we had two "ground" clocks only one million miles apart. As the astronaut, conveniently moving at $0.6c$, passed the first ground clock, both that clock and her own clock read zero.

As she passed the second ground clock, her own clock read 8 seconds and the *first* ground clock, which she photographed at that instant, she observed to read 4 seconds. That is to say, after 8 seconds had elapsed on her own clock, constant *observation* of the first ground clock would have revealed it to have registered only 4 seconds. (This effect is compounded of time dilation and the fact that as she moves away, the light from the clock is taking longer and longer to reach her.)

Our twin problem is the same thing, at the same speed, but over a longer time - we conclude that observation of any earth clock from the receding spacecraft will reveal it to be running *at half speed*, so the brother's flashes will be seen at the spacecraft to arrive every two months, by spacecraft time.

Symmetrically, as long as the brother on earth observes his sister's spacecraft to be moving away at $0.6c$, he will see light from her flashes to be arriving at the earth every two months by earth time. To figure the frequency of her brother's flashes observed as she returns towards earth, we have to go back to our previous example and find how the astronaut traveling at $0.6c$ observes time to be registered by the *second* ground clock, the one she's approaching.

We know that as she passes that clock, it reads 10 seconds and her own clock reads 8 seconds. We must figure out what she would have seen that second ground clock to read had she glanced at it through a telescope as she passed the first ground clock, at which point both her own clock and the first ground clock read zero. But at that instant, the reading she would see on the second ground clock must be the same as would be seen by an observer on the ground, standing by the first ground clock and observing the second ground clock through a telescope. Since the ground observer knows both ground clocks are synchronized, and the first ground clock reads zero, and the second is 6 light seconds distant, it must read -6 seconds if observed at that instant.

Hence the astronaut will observe the second ground clock to progress from -6 seconds to +10 seconds during the period that her own clock goes from 0 to 8 seconds. In other words, she sees the clock she is approaching at $0.6c$ to be running *at double speed.*

Finally, back to the twins. During her journey back to earth, the sister will see the brother's light flashing twice a month. (Evidently, the time dilation effect does not fully compensate for the fact that each succeeding flash has less far to go to reach her.)

We are now ready to do the bookkeeping: first, from the sister's point of view.

WHAT DOES SHE SEE?

At $0.6c$, she sees the distance to alpha-centauri to be contracted by the familiar $\sqrt{1-v^2/c^2} = 0.8$ to a distance of 3.2 light years, which at $0.6c$ will take her a time 5.333 years, or, more conveniently, 64 months. During the outward trip, then, she will see 32 flashes from home, she will see her brother to age by 32 months. Her return trip will also take 64 months, during which time she will see 128 flashes, so over the whole trip she will see

128 + 32 = 160 flashes. That means she will have seen her brother to age by 160 months or 13 years 4 months.

WHAT DOES HE SEE?

As he watches for flashes through his telescope, the stay-at-home brother will see his sister to be aging at half his own rate of aging as long as he sees her to be moving away from him, then aging at twice his rate as he sees her coming back. At first glance, this sounds the same as what she sees—but it isn't! The important question to ask is *when* does he see her turn around? To him, her outward journey of 4 light years' distance at a speed of 0.6*c* takes her 4/0.6 years, or 80 months. BUT he doesn't *see* her turn around until 4 years later, because of the time light takes to get back to earth from alpha-centauri. In other words, he will actually see her aging at half his rate for 80 + 48 = 128 months, during which time he will see 64 flashes.

When he *sees* his sister turn around, she is already more than half way back! Remember, in his frame the whole trip takes 160 months (8 light years at 0.6*c*) so he will only see her aging at twice his rate during the last 160 - 128 = 32 months, during which period he will see all 64 flashes she sent out on her return trip.

Therefore, by counting the flashes of light she transmitted once a month, he will conclude she has aged 128 months on the trip, which by his clock and calendar took 160 months. So when she steps off the spacecraft 32 months younger than her twin brother, neither of them will be surprised.

THE DOPPLER EFFECT

The above analysis hinges on the fact that a traveller approaching a flashing light at 0.6*c* will see it flashing at *double* its "natural" rate - the rate observed by someone standing still with the light - and a traveller receding at 0.6*c* from a flashing light will see it to flash at only *half* its natural rate.

This is a particular example of the *Doppler Effect*, first discussed in 1842 by the German physicist Christian Doppler. There is a Doppler Effect for sound waves too. Sound is generated by a vibrating object sending a succession of pressure pulses through the air. These pressure waves are analogous to the flashes of light. If you are approaching a sound source you will encounter the pressure waves more frequently than if you stand still. This means you will hear a higher frequency sound.

If the distance between you and the source of sound is increasing, you will hear a lower frequency. This is why the note of a jet plane or a siren goes lower as it passes you. The details of the Doppler Effect for sound are a little different than those for light, because the speed of sound is not the same for all observers - it's 330 meters per second relative to the air.

It isn't difficult to find the general formula for the Doppler shift, that is, the change in frequency observed when the source of waves (or periodic signals in general) is moving. For example, consider a light flashing once a second as observed by someone in the same frame as the light. Let us imagine the light to be attached to a spaceship, passing us at a relativistic speed v, and imagine we see a flash at the instant it passes us. When will we see the next flash? First, the spaceship's clock is running slow according to us, so it will take $1/\sqrt{1-v^2/c^2}$ seconds before it emits the next flash. But it's also moving away from us, so we won't see that next flash until the light has traveled back to us over the distance covered by the spaceship between flashes. Since the spaceship is traveling at v, and the time between flashes as measured in our frame is $1/\sqrt{1-v^2/c^2}$, the distance the spaceship covers between flashes measures in our frame is $v/\sqrt{1-v^2/c^2}$. Since the light coming back to us from the flash is traveling at c, it covers this distance in time $v/\sqrt{1-v^2/c^2}$.

Thus the total time between our observing the first flash as the spaceship passes close by us and the second flash emitted one second later by the spaceship clock is:

$$\frac{1}{\sqrt{1-v^2/c^2}}+\frac{v}{c\sqrt{1-v^2/c^2}}=\sqrt{\frac{1+v^2/c^2}{1-v/c}}$$

The change in frequency is the inverse of the change in time between flashes. Also, notice that the shift for a spaceship *approaching* at speed v is the *inverse* of that for one *receding* at speed v. We already established that above for the special case of $v = 0.6c$, where we found the frequency halved for the spaceship receding, doubled for it approaching.

There are some significant differences between the Doppler effect for light and that for sound. The most obvious one is that if something is approaching you at a speed higher than the speed of sound, you won't hear a thing until it hits. Nothing can approach you at greater than the speed of light. Another difference is that the Doppler shift for light depends only on the velocity of the emitter relative to that of the receiver. For sound, the shift depends on the velocities of both of them relative to the air. Finally, consider a signal from a moving object moving along a straight line, but not towards you.

For example, consider a train moving along a straight track, and the nearest point on the track is one mile from where you're standing. What frequency do you hear (or see) for the signal emitted when the train was at the nearest point to you on the track? For a sound signal, you would hear no Doppler shifting at this point. For a light signal, however, you *would* see a frequency shift downwards equal to the time dilation factor. This is called the *transverse* Doppler shift, and is famous historically because it was first detected in 1938 by two experimenters, Ives and Stillwell, who even then didn't believe in special relativity! They interpreted their result as the slowing down of a clock as it moved through the aether.

An important astronomical application of the Doppler Effect is the *red shift*. The light from very distant galaxies is redder than the light from similar galaxies nearer to us. This is because the further away a galaxy is, the faster it is moving away from us, as the Universe expands. The light is redder because red light is low frequency light (blue is high) and we see low frequency light for the same reason that the astronaut receding from earth sees flashes less frequently. In fact, the farthest away galaxies we can see are receding faster than the $0.6c$ of our astronaut.

ADDING VELOCITIES: A WALK ON THE TRAIN

THE FORMULA

If I walk from the back to the front of a train at 3 m.p.h., and the train is traveling at 60 m.p.h., then common sense tells me that my speed relative to the ground is 63 m.p.h. This obvious truth, the simple addition of velocities, follows from the Galilean transformations. Unfortunately, it can't be quite right for high speeds!

We know that for a flash of light going from the back of the train to the front, the speed of the light relative to the ground is exactly the same as its speed relative to the train, not 60 m.p.h. different. Hence it is necessary to do a careful analysis of a fairly speedy person moving from the back of the train to the front as viewed from the ground, to see how velocities *really* add.

We consider our standard train of length L moving down the track at steady speed v, and equipped with synchronized clocks at the back and the front.

The walker sets off from the back of the train when that clock reads zero. Assuming a steady walking speed of u meters per second (relative to the train, of course), the walker will see the front clock to read L/u seconds on arrival there.

How does this look from the ground? Let's assume that at the instant the walker began to walk from the clock at the back of the train, the back of the train was passing the ground observer's clock, and both these clocks (one on the train and one on the ground) read zero.

The ground observer sees the walker reach the clock at the front of the train at the instant that clock reads L/u (this is in agreement with what is observed *on* the train—two simultaneous events *at the same place* are simultaneous to all observers), but at this same instant, the ground observer says the train's *back* clock, where the walker began, reads $L/u + Lv/c^2$. (This follows from our previously established result that two clocks synchronized in one frame, in which they are L apart, will be out of synchronization in a frame in which they are moving at v along the line joining them by a time Lv/c^2.)

Now, how much time elapses as measured by the ground observer's clock during the walk? At the instant the walk began, the ground observer saw the clock at the back of the train (which was right next to him) to read zero. At the instant the walk ended, the ground observer would say that clock read $L/u + Lv/c^2$, from the paragraph above. But the ground observer would see that clock to be running slow, by the usual time dilation factor: so he would measure the time of the walk on his own clock to be:

$$\frac{L/u + Lv/c^2}{\sqrt{1-(v^2/c^2)}}.$$

How *far* does the walker move as viewed from the ground? In the time t_W, the train travels a distance vt_W, so the walker moves this distance plus the length of the train. Remember that the train is contracted as viewed from the ground! It follows that the distance covered relative to the ground during the walk is:

$$\begin{aligned} d_w &= vt_w + L\sqrt{1-(v^2/c^2)} \\ &= v\frac{L/u + Lv/c^2}{\sqrt{1-(v^2/c^2)}} + L\sqrt{1-(v^2/c^2)} \\ &= \frac{vL/u + Lv^2/c^2 + L - L(v^2 - c^2)}{\sqrt{1-(v^2/c^2)}} \\ &= \frac{L(1+v/u)}{\sqrt{1-(v^2/c^2)}} \end{aligned}$$

The walker's *speed* relative to the ground is simply d_W/t_W, easily found from the above expressions:

$$\frac{d_w}{t_w} = \frac{1+v/u}{1/u + v/c^2} = \frac{u+v}{1+uv/c^2}.$$

This is the appropriate formula for adding velocities. Note that it gives the correct answer, $u + v$, in the low velocity limit, and also if u or v equals c, the sum of the velocities is c.

Exercise: The direct derivation given above is really equivalent to a rederivation of the Lorentz equations. The equation describing the walker's path on the train is just $x' = ut'$. Substitute this in the Lorentz equations and prove it leads to a path relative to the ground given by $x = wt$, with w given by the velocity addition formula.

Exercise: Suppose a spaceship is equipped with a series of one-shot rockets, each of which can accelerate the ship to *c/2* from rest. It uses one rocket to leave the solar system (ignore gravity here) and is then traveling at $c/2$ (relative to us) in deep space. It now fires its second rocket, keeping the same direction. Find how fast it is moving relative to us. It now fires the third rocket, keeping the same direction. Find its new speed. Can you draw any general conclusions from your results?

WALKING *ACROSS* THE TRAIN

Imagine now a rather wide train, of width w, and the walker begins the walk across the train, which is now equipped with clocks on both sides, when the clock where he begins reads $t = 0$. For walking speed $u_{y'}$ (relative to the train, and across the train is the y-direction) when he reaches the clock at the other side it will read $w/u_{y'}$.

How is this seen from the ground? The width of the train w will be the same, there is no Lorentz contraction in the y-direction for motion in the x-direction. The beginning and ending clocks will also be synchronized as seen from the ground, since they are separated in the y-direction but not the x-direction. However, they are clocks moving at relativistic speed, so they will exhibit the familiar time dilation factor. That is, when they read $w/u_{y'}$, a clock on the ground will read

$$\frac{w}{u_{y'}}\frac{1}{\sqrt{1-v^2/c^2}}.$$

Thus, as observed from the ground, walking directly across the train is slowed down by the time dilation factor, just as is every other activity on the train as seen from the ground.

However, for steady motion on the train in an arbitrary direction, velocity components $(u_{x'}, u_{y'})$ the cross-train velocity transforms in a more complicated way, because the train clocks at the beginning and end of the walk are now separated in the x-direction, so if they register an elapsed time of w/u_y a ground observer would add a lack of synchronicity term

$$\frac{Lv}{c^2}=\frac{wu_{x'}}{u_{y'}}\frac{v}{c^2}.$$

Thus the time for the walk as observed from the ground

$$tw = \left(\frac{w}{u_{y'}} + \frac{wu_{x'}}{u_{y'}} \frac{v}{c^2} \right) / \sqrt{1 - v^2 / c^2} = \frac{w}{u_y}.$$

From this we find the general formula for transformation of transverse velocities:

$$u_y = u_{y'} \frac{\sqrt{1 - v^2 / c^2}}{\sqrt{1 + u_{x'} v / c^2}}.$$

For the special case of walking directly across the train, $u_{x'} = 0$, we recover the earlier result, that transverse velocity is simply slowed by the time dilation effect.

TESTING THE ADDITION OF VELOCITIES FORMULA

Actually, the first test of the addition of velocities formula was carried out in the 1850s! Two French physicists, Fizeau and Foucault, measured the speed of light in water, and found it to be c/n, where n is the refractive index of water, about 1.33. (This was the result predicted by the wave theory of light.)

They then measured the speed of light (relative to the ground) in *moving* water, by sending light down a long pipe with water flowing through it at speed v. They discovered that the speed relative to the ground was not just $v + c/n$, but had an extra term, $v + c/n - v/n^2$. Their (incorrect) explanation was that the light was a complicated combination of waves in the water and waves in the aether, and the moving water was only partially dragging the aether along with it, so the light didn't get the full speed v of the water added to its original speed c/n.

The true explanation of the extra term is much simpler: velocities don't simply add. To add the velocity v to the velocity c/n, we must use the addition of velocities formula above, which gives the light velocity relative to the ground to be:

$$(v + c/n)/(1 + v/nc)$$

Now, v is much smaller than c or c/n, so $1/(1 + v/nc)$ can be written as $(1 - v/nc)$, giving:

$$(v + c/n)(1 - v/nc)$$

Multiplying this out gives $v + c/n - v/n^2 - v/n \times v/c$, and the last term is smaller than v by a factor v/c, so is clearly negligible.

Therefore, the 1850 experiment looking for "aether drag" in fact confirms the relativistic addition of velocities formula! Of course, there are many other confirmations.

RELATIVISTIC DYNAMICS

The first coherent statement of what physicists now call *relativity* was Galileo's observation almost four hundred years ago that if you were in a large closed room, you could not tell by observing how things move-living things, thrown things, dripping liquids-whether the room was at rest in a building, say, or below decks in a large ship moving with a steady velocity.

More technically (but really saying the same thing!) we would put it that the laws of motion are the same in any inertial frame. That is, these laws really only describe relative positions and velocities. In particular, they do not single out a special inertial frame as the one that's "really at rest".

This was later all written down more formally, in terms of Galilean transformations. Using these simple linear equations, motion analyzed in terms of positions and velocities in one inertial frame could be translated into any other. When, after Galileo, Newton wrote down his Three Laws of Motion, they were of course invariant under the Galilean transformations, and valid in any inertial frame.

About two hundred years ago, it became clear that light was not just a stream of particles (as Newton had thought) but manifested definite *wavelike* properties.

This led naturally to the question of what, exactly, was waving, and the consensus was that space was filled with an *aether*, and light waves were ripples in this all-pervading aether analogous to sound waves in air. Maxwell's discovery that the equations describing electromagnetic phenomena had wavelike solutions, and predicted a speed which coincided with the measured speed of light, suggested that electric and magnetic fields were stresses or strains in the aether, and Maxwell's equations were presumably only precisely correct in the frame in which the aether was at rest. However, very precise experiments which should have been able to detect this aether all failed.

About a hundred years ago, Einstein suggested that maybe *all* the laws of physics were the same in *all* inertial frames, generalizing Galileo's pronouncements concerning motion to include the more recently discovered laws of electricity and magnetism. This would imply there could be no special "really at rest" frame, even for light propagation, and hence no aether. This is a very appealing and very simple concept: the same laws apply in all frames.

What could be more reasonable? Though, it turns out to clash with some beliefs about space and time deeply held by everybody encountering this for the first time. The central prediction is that since the speed of light follows from the laws of physics (Maxwell's equations) and some simple electrostatic and magnetostatic experiments, which are clearly frame-independent, the speed of light is the same in all inertial frames.

That is to say, the speed of a particular flash of light will always be measured to be $3{,}10^8$ meters per second even if measured by different observers moving rapidly relative to each other, where each observer measures the speed of the flash relative to himself. Nevertheless, experiments have show again and again that Einstein's elegant insight is right, and everybody's deeply held beliefs are wrong.

We have discussed in detail the *kinematical* consequences of Einstein's postulate: how measurements of position, time and velocity in one frame relate to those in another, and how apparent paradoxes can be resolved by careful analysis.

So far, though, we have not thought much about dynamics. We know that Newton's Laws of Motion were invariant under the Galilean transformations between inertial frames. We now know that the Galilean transformations are in fact *incorrect* except in the low speed nonrelativistic limit. Therefore, we had better look carefully at Newton's Laws of Motion in light of our new knowledge.

NEWTON'S LAWS REVISITED

Newton's First Law, the Principle of Inertia, that an object subject to no external forces will continue to move in a straight line at steady speed, is equally valid in special relativity. Indeed, it is the defining property of an inertial frame that this *is* true, and the content of special relativity is transformations between such frames.

Newton's Second Law, stated in the form force = mass x acceleration, cannot be true as it stands in special relativity. This is evident from the formula we derived for addition of velocities. Think of a rocket having many stages, each sufficient to boost the remainder of the rocket (including the unused stages) to *c/2* from rest. We could fire them one after the other in a carefully timed way to generate a continuous large force on the rocket, which would get it to *c/2* in the first firing. If the acceleration continued, the rocket would very soon be exceeding the speed of light. Yet we know from the addition of velocities formula that in fact the rocket never reaches *c*.

Evidently, Newton's Second Law needs updating. Newton's Third Law, action = reaction, also has problems. Consider some attractive force between two rapidly moving bodies. As their distance apart varies, so does the force of attraction. We might be tempted to say that the force of *A* on *B* is the opposite of the force of *B* on *A*, at each instant of time, but that implies *simultaneous* measurements at two bodies some distance from each other, and if it happens to be true in *A*'s inertial frame, it won't be in *B*'s.

CONSERVATION LAWS

In nonrelativistic Newtonian physics, the Third Law tells us that two interacting bodies feel equal but opposite forces from the interaction.

Therefore from the Second Law, the rate of change of momentum of one of the bodies is equal and opposite to that of the other body, thus the total rate of change of momentum of the system caused by the interaction is zero. Consequently, for any closed dynamical system (no outside forces acting) the total momentum never changes. This is the law of conservation of momentum. It does not depend on the details of the forces of interaction between the bodies, only that they be equal and opposite. The other major dynamical conservation law is the conservation of energy.

This was not fully formulated until long after Newton, when

it became clear that frictional heat generation, for example, could quantitatively account for the apparent loss of kinetic plus potential energy in actual dynamical systems.

Although these conservation laws were originally formulated within a Newtonian worldview, their very general nature suggested to Einstein that they might have a wider validity. Therefore, as a working hypothesis, he assumed them to be satisfied *in all inertial frames*, and explored the consequences. We follow that approach.

MOMENTUM CONSERVATION ON THE POOL TABLE

Let us consider conservation of momentum for a collision of two balls on a pool table. We draw a chalk line down the middle of the pool table, and shoot the balls close to, but on opposite sides of, the chalk line from either end, at the same speed, so they will hit in the middle with a glancing blow, which will turn their velocities through a small angle. In other words, if initially we say their (equal magnitude, opposite direction) velocities were parallel to the x-direction—the chalk line—then after the collision they will also have equal and opposite small velocities in the y-direction. (The x-direction velocities will have decreased very slightly).

A SYMMETRICAL SPACESHIP COLLISION

Now let us repeat the exercise on a grand scale. Suppose somewhere in space, far from any gravitational fields, we set out a string one million miles long. (It could be between our two clocks in the time dilation experiment).

This string corresponds to the chalk line on the pool table. Suppose now we have two identical spaceships approaching each other with equal and opposite velocities parallel to the string from the two ends of the string, aimed so that they suffer a slight glancing collision when they meet in the middle. It is evident from the symmetry of the situation that momentum is conserved in both directions. In particular, the rate at which one spaceship moves away from the string after the collision - its y-velocity - is equal and opposite to the rate at which the other one moves away from the string.

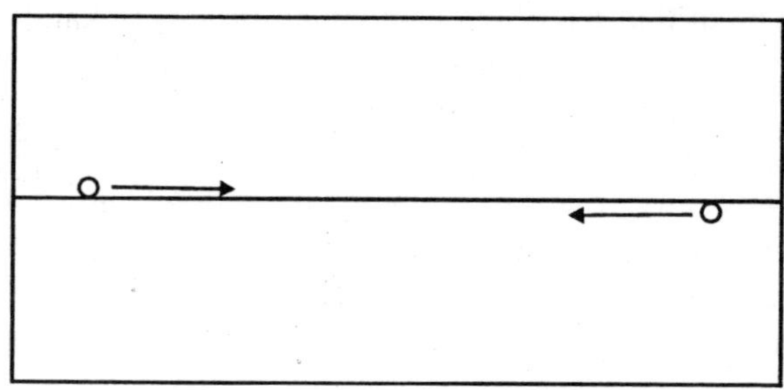

Balls on pool table moving towards glancing collision

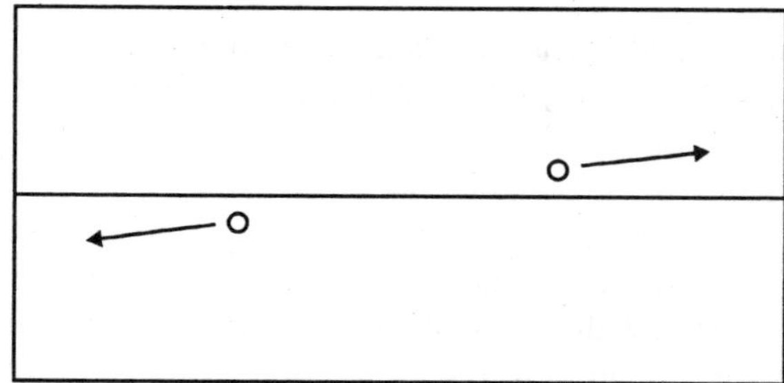

Motion of balls on table after collision

But now consider this collision as observed by someone in one of the spaceships, call it *A*. (Remember, momentum must be conserved in *all* inertial frames—they are all equivalent—there is nothing special about the frame in which the string is at rest.) Before the collision, he sees the string moving very fast by the window, say a few meters away.

After the collision, he sees the string to be moving away, at, say, 15 meters per second. This is because spaceship *A* has picked up a velocity perpendicular to the string of 15 meters per second. Meanwhile, since this is a completely symmetrical situation, an observer on spaceship *B* would certainly deduce that her spaceship was moving away from the string at 15 meters per second as well.

JUST HOW SYMMETRICAL IS IT?

The crucial question is: how fast does an observer in spaceship A see spaceship B to be moving away from the string? Let us suppose that relative to spaceship A, spaceship B is moving away (in the x-direction) at 0.6c.

First, recall that distances perpendicular to the direction of motion are not Lorentz contracted. Therefore, when the observer in spaceship *B* says she has moved 15 meters further away from the string in a one second interval, the observer watching this movement from spaceship *A* will agree on the 15 meters - but disagree on the one second! He will say her clocks run slow, so as measured by his clocks 1.25 seconds will have elapsed as she moves 15 meters in the *y*-direction.

It follows that, as a result of time dilation, this collision as viewed from spaceship *A* does *not* cause equal and opposite velocities for the two spaceships in the *y*-direction. Initially, both spaceships were moving parallel to the *x*-axis - there was zero momentum in the *y*-direction.

Consider *y*-direction momentum conservation in the inertial frame in which *A* was initially at rest. An observer in that frame measuring *y*-velocities after the collision will find *A* to be moving at 15 meters per second, *B* to be moving at -0.8 x 15 meters per second in the *y*-direction. So how can we argue there is zero total momentum in the *y*-direction *after* the collision, when the identical spaceships do *not* have equal and opposite velocities?

EINSTEIN RESCUES MOMENTUM CONSERVATION

Einstein was so sure that momentum conservation must always hold that he rescued it with a bold hypothesis: the mass of an object must depend on its speed! In fact, the mass must increase with speed in just such a way as to cancel out the lower *y*-direction velocity resulting from time dilation. That is to say, if an object at rest has a mass m_0, moving at a speed v it must have mass

$$m = \frac{m_0}{\sqrt{1 - v^2 / c^2}}$$

to conserve *y*-direction momentum.

Note that this is an undetectably small effect at ordinary speeds, but as an object approaches the speed of light, the mass increases without limit!

Of course, we have taken a very special case here: a particular kind of collision. The reader might well wonder if the same mass correction would work in other types of collision, for example a straight line collision in which a heavy object rear-ends a lighter object. The algebra is straightforward, if tedious, and it is found that this mass correction factor does indeed ensure momentum conservation for any collision in all inertial frames.

MASS REALLY *DOES* INCREASE WITH SPEED

Deciding that masses of objects must depend on speed like this seems a heavy price to pay to rescue conservation of momentum! However, it is a prediction that is not difficult to check by experiment. The first confirmation came in 1908, measuring the mass of fast electrons in a vacuum tube. In fact, the electrons in an old-fashioned colour TV tube are about half a percent heavier than electrons at rest, and this must be allowed for in calculating the magnetic fields used to guide them to the screen.

Much more dramatically, in modern particle accelerators very powerful electric fields are used to accelerate electrons, protons and other particles. It is found in practice that these particles become heavier and heavier as the speed of light is approached, and hence need greater and greater forces for further acceleration. Consequently, the speed of light is a natural absolute speed limit. Particles are accelerated to speeds where their mass is thousands of times greater than their mass measured at rest, usually called the "rest mass".

Warning: It should be mentioned that some people don't like the statement that mass increases with speed, they feel that the word "mass" should be restricted to the rest mass of an object, which we've called m_0. This difference of definition has no physical content, however—it's just a matter of taste. We would write momentum as $p = mv$, they would write our m_0 as m, and say the formula for momentum in their notation is $p = mv/\sqrt{1-v^2/c^2}$. Either way, a fast electron is that much harder to deflect from a straight line.

MASS AND ENERGY CONSERVATION: KINETIC ENERGY AND MASS FOR VERY FAST PARTICLES

As everyone has heard, in special relativity mass and energy are not separately conserved, in certain situations mass m can be converted to energy $E = mc^2$. This equivalence is closely related to the mass increase with speed. Suppose a constant force F accelerates a particle of rest mass m_0 in a straight line. The work done by the force in accelerating the particle as it travels a distance d is Fd, and this work has given the particle kinetic energy.

The elementary derivation of the kinetic energy $\frac{1}{2}mv^2$ of an ordinary *non-relativistic* (i.e. slow moving) object of mass m. Suppose it starts from rest. Then after time t, it has traveled distance $d = \frac{1}{2} at^2$, and $v = at$. From Newton's second law, $F = ma$, the work done by the force $Fd = mad = \frac{1}{2} ma^2t^2 = \frac{1}{2} mv^2$.

This won't work if the mass is varying, because Newton's Second Law isn't always $F = ma$, for variable mass it's

$$F = dp/dt,$$

force = rate of change of momentum, and if the mass changes the momentum changes, even at constant velocity.

An instructive extreme case is the kinetic energy of a particle traveling close to the speed of light, as particles do in accelerators. In this regime, the change of speed with increasing momentum is negligible! Instead,

$$F = \frac{dp}{dt} = \frac{d(mv)}{dt} \cong \frac{dm}{dt}c$$

where as usual c is the speed of light. This is what happens in a particle accelerator for a charged particle in a constant electric field, with $F = qE$.

Since the particle is moving at a speed very close to c, in time dt it will move cdt and the force will do work $Fcdt$. The equation above can be rewritten

$$Fcdt = (dm)c^2$$

So the energy dE expended by the accelerating force in the time dt yields an increase in mass, and $dE = (dm)c^2$. Provided the speed is close to c, this can of course be integrated to an excellent approximation, to relate a finite particle mass change to the energy expended in accelerating it.

KINETIC ENERGY AND MASS FOR SLOW PARTICLES

Recall that to get momentum to be conserved in all inertial

frames, we had to assume an increase of mass with speed by the factor $1/\sqrt{1-v^2/c^2}$. This necessarily implies that even a slow-moving object has a tiny mass increase if it is put in motion.

How does this mass increase relate to the kinetic energy? Consider a mass m_0, moving at speed v, much less than the speed of light. Its kinetic energy $E=\frac{1}{2}m_0v^2$, as discussed above. Its mass is $m_0/\sqrt{1-v^2/c^2}$ which we can write as $m_0 + dm$, so dm is the tiny mass increase we know must occur. It's easy to calculate dm.

For $v/c \ll 1$, we can make the approximations

$$\sqrt{1-v^2/c^2} \cong 1-\frac{1}{2}v^2/c^2$$

and

$$\frac{1}{1-\frac{1}{2}v^2/c^2} \cong 1+\frac{1}{2}v^2/c^2.$$

So, for $v/c \ll 1$,

$$m(v) \cong m_0\left(1+\frac{1}{2}v^2/c^2\right)$$

$$dm \cong \left(\frac{1}{2}m_0v^2\right)/c^2 = KE/c^2.$$

Again, the mass increase dm is related to the kinetic energy KE by $KE = (dm)c^2$. Having looked at two simple cases, we're ready to derive the general result, valid over the whole range of possible speeds.

KINETIC ENERGY AND MASS FOR PARTICLES OF ARBITRARY SPEED

We have shown in the two sections above that (in the two limiting cases) when a force does work to increase the kinetic energy of a particle it also causes the mass of the particle to increase by an amount equal to the increase in energy divided by c^2.

In fact this result is exactly true over the whole range of speed from zero to arbitrarily close to the speed of light. For a particle of rest mass m_0 accelerating along a straight line (from rest) under a constant force F,

$$F = \frac{d}{dt}(mv)$$

$$= \frac{dm}{dt}v + m\frac{dv}{dt}$$

$$= \frac{m_0}{\left(1/v^2/c^2\right)^{3/2}}\frac{v^2}{c^2} + \frac{m_0}{\left(1/v^2/c^2\right)^{1/2}}\frac{dv}{dt}$$

$$= \frac{m_0}{\left(1/v^2/c^2\right)^{3/2}}\frac{dv}{dt}.$$

Therefore, the work done when the particle moves a distance dx is

$$Fdx = \frac{m_0}{\left(1/v^2/c^2\right)^{3/2}}\frac{dv}{dt}dx$$

$$= \frac{m_0}{\left(1/v^2/c^2\right)^{3/2}}vdv,$$

using $v = dx/dt$.

$$\int Fdx = \int \frac{m_0}{\left(1/v^2/c^2\right)^{3/2}}vdv = (m - m_0)c^2.$$

Therefore the total work done from rest—the kinetic energy—is:

(The integral is easily done by making the substitution $y = v^2/c^2$.)

So we see that in the general case the work done on the body, by definition its kinetic energy, is just equal to its mass increase multiplied by c^2.

To understand why this isn't noticed in everyday life, try an example, such as a jet airplane weighing 100 tons moving at

2,000mph. 100 tons is 100,000 kilograms, 2,000mph is about 1,000 meters per second. That's a kinetic energy ½ mv^2 of ½.10^{11}joules, but the corresponding mass change of the airplane down by the factor $c^2 = 9.10^{16}$, giving an actual mass increase of about half a milligram, not too easy to detect!

Notation m and m_0: As stated earlier, we use m_0 to denote the "rest mass" of an object, and m to denote its relativistic mass $m = m_0 / \sqrt{1 - v^2 / c^2}$,

In this notation, we follow French and Feynman. *Krane and Tipler, in contrast, use m for the rest mass.* Using m as we do gives neater formulas for momentum and energy, but is not without its dangers. One must remember that m is *not* a constant, but a function of speed. Also, *one must remember* that the relativistic kinetic energy is $(m-m_0)c^2$, and *not* equal to ½mv^2, even with the relativistic mass!

Example: take $v^2/c^2 = 0.99$, find the kinetic energy, and compare it with ½mv^2 (using the relativistic mass).

Chapter 9

Mass and Energy

REST ENERGY

The fact that feeding energy into a body increases its mass suggests that the mass m_0 of a body *at rest*, multiplied by c^2, can be considered as a quantity of energy. The truth of this is best seen in interactions between elementary particles. For example, there is a particle called a *positron* which is exactly like an electron except that it has positive charge. If a positron and an electron collide at low speed (so there is very little kinetic energy) they both disappear in a flash of electromagnetic radiation. This can be detected and its energy measured. It turns out to be $2m_0c^2$ where m_0 is the mass of the electron (and the positron).

Thus particles can "vaporize" into pure energy, that is, electromagnetic radiation. The energy m_0c^2 of a particle at rest is called its "rest energy". Note, however, that an electron can only be vaporized by meeting with a positron, and there are very few positrons around normally, for obvious reasons-they just don't get far. (Although occasionally it has been suggested that some galaxies may be antimatter!)

EINSTEIN'S BOX

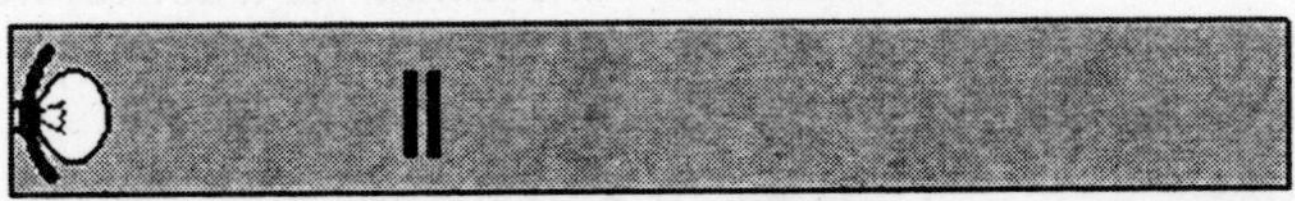

An amusing "experiment" on the equivalence of mass and energy is the following: consider a closed box with a flashlight at one end and light-absorbing material at the other end. Imagine the box to be far out in space away from gravitational fields or any disturbances. Suppose the light flashes once, the flash travels down the box and is absorbed at the other end.

Einstein's Box: far out in space, a flash of light emitted by a bulb attached at one end is absorbed at the other end. Now it is known from Maxwell's theory of electromagnetic waves that a flash of light carrying energy E also carries momentum $p = E/c$. Thus, as the flash leaves the bulb and goes down the tube, the box recoils, like a gun, to conserve overall momentum. Suppose the whole apparatus has mass M and recoils at velocity v. Of course, $v << c$. Then from conservation of momentum in the frame in which the box was initially at rest:

$$Mv = E/c$$

the recoil momentum of the box equals (minus) the momentum of the flash emitted. After a time $t = L/c$ the light hits the far end of the tube, is absorbed, and the whole thing comes to rest again. (We are assuming that the distance moved by the box is tiny compared to its length.) How far did the box move? It moved at speed v for time t, so it moved distance

$$d = vt = vL/c$$

From the conservation of momentum equation above, we see that $v = E/Mc$, so the distance d the box moved over is:

$$d = \frac{EL}{mc^2}.$$

Now, the important thing is that there are *no* external forces acting on this system, so the centre of mass cannot have moved!

The only way this makes sense is to say that to counterbalance the mass M moving d backwards, the light energy must have transferred a small mass m, say, the length L of the tube so that

$$Md = mL$$

and balance is maintained. From our formula for d above, we can figure out the necessary value of m,

$$m = \frac{M}{L}d = \frac{M}{L}\frac{EL}{Mc^2} = \frac{E}{c^2}$$

so

$$E = mc^2.$$

We have therefore established that transfer of energy implies transfer of the equivalent mass. Our only assumptions here are that the centre of mass of an isolated system, initially at rest, remains at rest if no external forces act, and that electromagnetic radiation carries momentum E/c, as predicted by Maxwell's equations and experimentally established.

But how is this mass transfer physically realized? Is the front end of the tube really heavier after it absorbs the light? The answer is yes, because it's a bit hotter, which means its atoms are vibrating slightly faster—and faster moving objects have higher mass. (And there's another contribution we're about to discuss.)

MASS AND POTENTIAL ENERGY

Suppose now at the far end of the tube we have a hydrogen atom at rest. This atom is essentially a proton having an electron bound to it by electrostatic attraction. It is known that a flash of light with total energy 13.6eV is just enough to tear the electron away, so in the end the proton and electron are at rest far away from each other.

The energy of the light was used up dragging the proton and electron apart—that is, it went into potential energy. (It should be mentioned that the electron also loses kinetic energy in this process, 13.6 ev is the net energy required to break up the atom.) Now, the light is absorbed by this process, so from our argument above the right hand end of the tube must become heavier. That is to say, a proton at rest plus a (distant) electron at rest weigh more than a hydrogen atom by E/c^2, with E equal to 13.6eV. Thus, Einstein's box forces us to conclude that increased *potential* energy in a system also entails the appropriate increase in mass.

It is interesting to consider the hydrogen atom dissociation in reverse—if a slow moving electron encounters an isolated proton, they may combine to form a hydrogen atom, emitting 13.6eV of electromagnetic radiation energy as they do so. Clearly, then, the hydrogen atom remaining has that much less energy than the initial proton + electron.

The actual mass difference for hydrogen atoms is about one part in 10^8. This is typical of the energy radiated away in a violent chemical reaction—in fact, since most atoms are an order of magnitude or more heavier than hydrogen, a part in 10^9 or 10^{10} is more usual. However, things are very different in nuclear physics, where the forces are stronger so the binding is tighter. We shall discuss this later, but briefly mention an example: a hydrogen nucleus can combine with a lithium nucleus to give two helium nuclei, and the mass shed is 1/500 of the original. This reaction has been observed, and all the masses involved are measurable. The actual energy emitted is 17 MeV. This is the type of reaction that occurs in hydrogen bombs. Notice that the energy released is at least a million times more than the most violent chemical reaction.

As a final example, let's make a ballpark estimate of the change in mass of a million tons of TNT on exploding. The TNT molecule is about a hundred times heavier than the hydrogen atom, and gives off a few eV on burning. So the change in weight is of order 10^{-10} x10^6 tons, about a hundred grams. In a hydrogen bomb, this same mass to energy conversion would take about fifty kilograms of fuel.

FOOTNOTE: EINSTEIN'S BOX IS A FAKE

Although Einstein's box argument is easy to understand, and gives the correct result, it is based on a physical fiction—the rigid box. If we had a rigid box, or even a rigid stick, all our clock synchronization problems would be over—we could start clocks at the two ends of the stick simultaneously by nudging the stick from one end, and, since it's a rigid stick, the other end would move instantaneously.

Actually there are no such materials. All materials are held together by electromagnetic forces, and pushing one end causes a wave of compression to travel down the stick. The electrical forces between atoms adjust at the speed of light, but the overall wave travels far more slowly because each atom in the chain must accelerate for a while before it moves sufficiently to affect the next one measurably.

So the light pulse will reach the other end of the box before it has begun to move! Nevertheless, the wobbling elastic box *does* have a net recoil momentum, which it does lose when the light hits the far end. So the basic point is still valid. French gives a legitimate derivation, replacing the box by its two (disconnected) ends, and finding the centre of mass of this complete system, which of course remains at rest throughout the process.

ENERGY AND MOMENTUM IN LORENTZ TRANSFORMATIONS

TOTAL ENERGY OF A PARTICLE DEPEND ON SPEED

We have a formula for the total energy E = K.E. + rest energy,

$$E = mc^2 = \frac{m_0 c^2}{\sqrt{1 - v^2 / c^2}}$$

so we can see how total energy varies with speed.

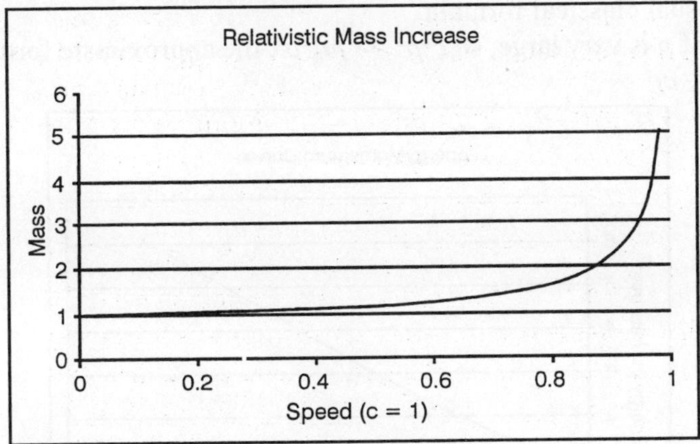

The momentum varies with speed as

$$p = mv = \frac{m_0 v}{\sqrt{1 - v^2 / c^2}}$$

TOTAL ENERGY OF A PARTICLE DEPEND ON MOMENTUM

It turns out to be useful to have a formula for E in terms of p. Now

$$E^2 = m^2 c^4 = \frac{m_0^2 c^4}{1 - v^2 / c^2}$$

so

$$m^2c^4 (1 - v^2/c^2) = m_0^2 c^4$$

$$m^2c^4 - m^2v^2c^2 = m_0^2 c^4$$

$$m^2c^4 = E^2 = m_0^2 c^4 + m^2c^2v^2$$

hence using $p = mv$ we find

$$E = \sqrt{m_0^2 c^4 + c^2 p^2} \, .$$

If p is very small, this gives

$$E \approx m_0 c^2 + \frac{p^2}{2m_0} ,$$

the usual classical formula.

If p is very large, so $c^2p^2 >> m_0{}^2c^4$, the approximate formula is $E = cp$.

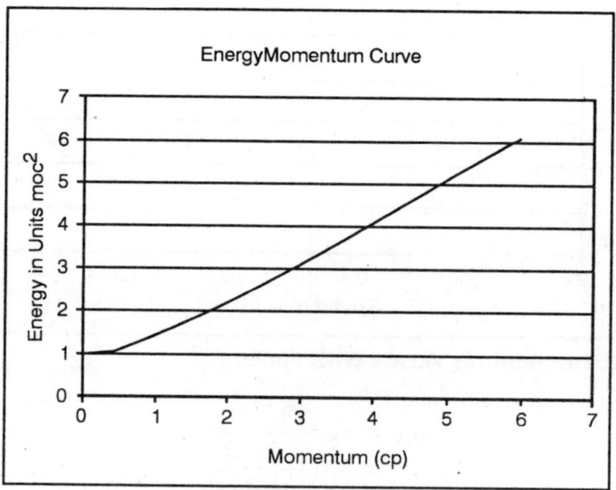

THE HIGH KINETIC ENERGY LIMIT: REST MASS BECOMES UNIMPORTANT!

Notice that this high energy limit is just the energy-momentum relationship Maxwell found to be true for light, for all p. This could only be true for *all* p if $m_0{}^2c^4 = 0$, that is, $m_0 = 0$.

Light is in fact composed of "photons"—particles having zero "rest mass".

The "rest mass" of a photon is meaningless, since they're never at rest—the energy of a photon

$$E^2 = mc^2 = \frac{m_0c^2}{\sqrt{1-v^2/c^2}}$$

is of the form 0/0, since $m_0 = 0$ and $v = c$, so "m" can still be nonzero. That is to say, the mass of a photon is really all K.E. mass.

For very fast electrons, such as those produced in high energy accelerators, the additional K.E. mass can be thousands of times the rest mass. For these particles, we can neglect the rest mass and take $E = cp$.

TRANSFORMING ENERGY AND MOMENTUM TO A NEW FRAME

We have shown

$$\vec{p} = m\vec{v} = \frac{m_0\vec{v}}{\sqrt{1-v^2/c^2}}$$

$$E^2 = mc^2 = \sqrt{m_0^2c^4 + c^2 - \vec{p}^2}\,.$$

Notice we can write this last equation in the form

$$E^2 - c^2\vec{p}^2 = m_0^2c^4\,.$$

That is to say, $E^2 - c^2\vec{p}^2$ depends *only* on the rest mass of the particle and the speed of light. It does not depend on the velocity of the particle, so it must be the same—for a particular particle—in all inertial frames.

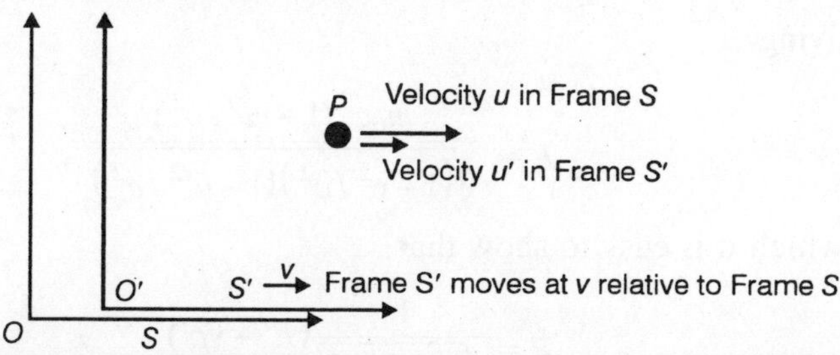

This is reminiscent of the invariance of $\vec{x}^2 - c^2t^2$, the interval squared between two events, under the Lorentz transformations. One might guess from this that the laws governing the transformation from E, p in one Lorentz frame to E', p' in another are similar to those for x,t. We can actually derive the laws for E, p to check this out.

As usual, we consider all velocities to be parallel to the x-axis.

We take the frame S' to be moving in the x-direction at speed v relative to S.

Consider a particle of mass m_0 (rest mass) moving at u' in the x' direction in frame S', and hence at u along x in S, where

$$u = \frac{u' + v}{1 + vu'/c^2}$$

The energy and momentum in S' are

$$E' = \frac{m_0 c^2}{\sqrt{1 - u'^2/c^2}}, p' = \frac{m_0 u'}{\sqrt{1 - u'^2/c^2}}.$$

and in S:

$$E = \frac{m_0 c^2}{\sqrt{1 - u^2/c^2}}, p = \frac{m_0 u}{\sqrt{1 - u^2/c^2}}$$

Thus

$$E = \frac{m_0 c^2}{\sqrt{1 - \left(\frac{u' + v}{1 + vu'/c^2}\right)^2 / c^2}}$$

giving

$$E = \frac{m_0 c^2 (1 + vu'/c^2)}{\sqrt{(1 - v^2/c^2)(1 - u'^2/c^2)}}$$

from which it is easy to show that

$$E = \frac{1}{\sqrt{1 - v^2/c^2}} (E' + vp').$$

Similarly, we can show that

$$p = \frac{p' + vE'/c^2}{\sqrt{1 - v^2/c^2}}$$

These are the Lorentz transformations for energy and momentum of a particle—it is easy to check that

$$E^2 - c^2p^2 = E'^2 - c^2 p'^2 = m_0^2 c^4 .$$

PHOTON ENERGIES IN DIFFERENT FRAMES

For a zero rest mass particle, such as a photon, $E = cp$, $E^2 - c^2p^2 = 0$ in all frames.

Thus

$$E = \frac{E' + vp'}{\sqrt{1 - v^2/c^2}} = \frac{E' + vE'/c}{\sqrt{1 - v^2/c^2}} = E'\sqrt{\frac{1 + v/c}{1 - v/c}}$$

Since $E = cp$, $E' = cp'$ we also have

$$p = p'\sqrt{\frac{1 + v/c}{1 - v/c}} .$$

Notice that the ratios of photon *energies* in the two frames coincides with the ratio of photon *frequencies* found in the Doppler shift. The photon energy is proportional to the frequency, so these two must of course transform in identical fashion.

But it's interesting to see it come about this way. Needless to say, relativity gives us no clue on what the constant of proportionality (Planck's constant) is: it must be measured experimentally. But the *same* constant plays a role in all quantum phenomena, not just those concerned with photons.

TRANSFORMING ENERGY INTO MASS: PARTICLE CREATION

PION PRODUCTION

We have mentioned how, using a synchrocyclotron, it is possible to accelerate protons to relativistic speeds. The rest energy of a proton m_pc^2 is 938 MeV, using here the standard high energy physics energy unit: 1 MeV = 10^6 eV. The neutron is a bit heavier—m_nc^2 = 940 MeV. (The electron is 0.51 MeV). Thus to accelerate a proton to relativistic speeds implies giving it a K.E. of order 1,000 MeV, or 1 GeV.

The standard operating procedure of high energy physicists is to accelerate particles to relativistic speeds, then smash them into other particles to see what happens. For example, fast protons will be aimed at protons at rest (hydrogen atoms, in other words—the electron can be neglected).

These proton-proton collisions take place inside some kind of detection apparatus, so the results can be observed. One widely-used detector is the bubble chamber: a transparent container filled with a superheated liquid. The electric field of a rapidly moving charged particle passing close to a molecule can dislodge an electron, so an energetic particle moving through the liquid leaves a trail of ionized molecules. These give centers about which bubbles can nucleate. The bubbles grow rapidly and provide a visible record of the particle's path.

What is actually observed in *p-p* scattering at relativistic energies is that often more particles come out than went in—particles called pions, p^+, p^0, p^- can be created. The p^0 is electrically neutral, the p^+ has exactly the same amount of charge as the proton. It is found experimentally that total electric charge is always conserved in collisions, no matter how many new particles are spawned, and total baryon number (protons + neutrons) is conserved.

Possible scenarios include:

$$p + p \rightarrow p + p + \pi^0$$

and

$$p + p \rightarrow p + n + \pi^+.$$

The neutral pion mass is 135 MeV, the charged pions have mass 140 MeV, where we follow standard high energy practice in calling mc^2 the "mass", since this is the energy equivalent, and hence the energy which, on creation of the particle in a collision, is taken from kinetic energy and stored in mass.

However, an incoming proton with 135 MeV of kinetic energy will not be able to create a neutral pion in a collision with a stationary proton. This is because the incoming proton also has momentum, and the collision conserves momentum, so some of the particles after the collision must have momentum and hence kinetic energy.

The simplest way to figure out just how much energy the incoming proton needs to create a neutral pion is to go to the centre of mass frame, where initially two protons are moving towards each other with equal and opposite velocities, there is no total momentum. Obviously, in this frame the least possible K.E. must be just enough to create the p^0 with all the final state particles (p,p,p^0) at rest. Thus if the relativistic mass of the incoming protons in the centre of mass frame is m, the total energy

$$E = 2mc^2 = 2m_y c^2 + m_z c^2.$$

Putting in the proton and pion masses from above, and using $m = m_p / \sqrt{1-(v^2/c^2)}$, we find the two incoming protons must both be traveling at $0.36c$.

Recall that this is the speed in the centre of mass frame, and for practical purposes, like designing the accelerator, we need to know the energy necessary in the "lab" frame—that in which one of the protons is initially at rest. The two frames obviously have a relative speed of $0.36c$, so to get the speed of the incoming proton in the lab frame we must add a velocity of $0.36c$ to one of $0.36c$ using the relativistic addition of velocities formula, which gives $0.64c$. This implies the incoming proton has a relativistic mass of 1.3 times its rest mass, and thus a K.E. around 280 MeV.

Thus to create a pion of rest energy 135 MeV, it is necessary to give the incoming proton at least 290 MeV of kinetic energy. This is called the "threshold energy" for pion production. The "inefficiency" arises because momentum is also conserved, so there is still considerable K.E. in the final particles.

ANTIPROTON PRODUCTION

On raising the energy of the incoming proton further, more particles are produced, including the "antiproton"—a negatively charged heavy particle which will annihilate a proton in a flash of energy. It turns out experimentally that an antiproton can only be produced accompanied by a newly created proton,

$$p + p \rightarrow p + p + p + \overline{p}$$

Notice we could have conserved electric charge with less energy with the reaction

$$p + p \rightarrow p + p + \pi^+ + \overline{p}$$

but this doesn't happen—so charge conservation isn't the only constraint on which particles can be produced. In fact, what we are seeing here is experimental confirmation that the conservation of baryon number, which at the low energies previously discussed in the context of pion production just meant that the total number of protons plus neutrons stayed fixed, is generalized at high energies to include antiparticles with negative baryon number, –1 for the antiproton.

Thus baryon number conservation becomes parallel to electric charge conservation, new particles can always be produced at high enough energies provided the total new charge and the total new baryon number are both zero. (Actually there are further conservation laws which become important when more exotic particles are produced, which we may discuss later.) We should emphasize again that these are *experimental* results gathered from examining millions of collisions between relativistic particles.

One of the first modern accelerators, built at Berkeley in the fifties, was designed specifically to produce the antiproton, so it was very important to calculate the antiproton production threshold correctly! This can be done by the same method we used above for pion production, but we use a different trick here which is often useful. We have shown that on transforming the energy and momentum of a particle from one frame to another

$$E^2 - c^2 p^2 = E'^2 - c^2 p'^2.$$

Since the Lorentz equations are linear, if we have a system of particles with total energy E and total momentum p in one frame, E', p' in another, it must again be true that

$$E^2 = c^2 \vec{p}^2 = E'^2 - c^2 \vec{p}'^2.$$

We can use this invariance to get lab frame information from the centre of mass frame. Noting that in the centre of mass (CM) frame the momentum is zero, and in the lab frame the momentum is all in the incoming proton,

$$E_{cm}^2 = ((m_{in} + m_0)c^2)^2 - c^2 p_{in}^2$$

where here m_0 is the proton rest mass, m_{in} is the relativistic mass of the incoming proton. At the antiproton production threshold,

$$E_{cm} = 4m_0c^2, \text{ so}$$

$$16m_0^2c^4 = m_{in}^2c^4 + 2m_{in}c^2m_0c^2 + m_0^2c^4 - c^2p_{in}^2$$

and using

$$m_{in}^2c^4 - c^2p_{in}^2 = m_0c^4$$

we find

$$2(m_{in}c^2)(m_0c^2) + 2(m_0c^2)^2 = 16(m_0c^2)^2,$$

so

$$m_{in}c^2 = 7m_0c^2.$$

Therefore to create two extra particles, with total rest energy $2m_0c^2$, it is necessary for the incoming proton to have a kinetic energy of $6m_0c^2$. The Berkeley Gevatron had design energy 6.2 GeV.

As we go to higher energies, this "inefficiency" gets worse—consider energies such that the kinetic energy >> rest energy, and assume the incoming particle and the target particle have the same rest mass, m_0, with the incoming particle having relativistic mass m.

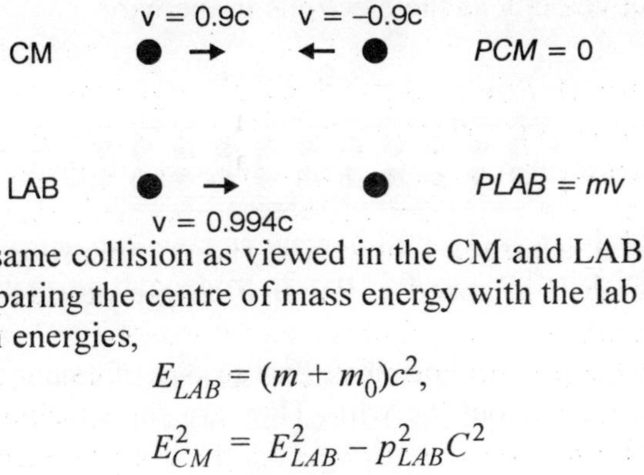

	v = 0.9c	v = –0.9c	
CM	● →	← ●	PCM = 0
LAB	● →	●	PLAB = mv
	v = 0.994c		

The same collision as viewed in the CM and LAB frames.

Comparing the centre of mass energy with the lab energy at these high energies,

$$E_{LAB} = (m + m_0)c^2,$$

$$E_{CM}^2 = E_{LAB}^2 - p_{LAB}^2C^2$$

$$= m^2c^4 + 2mc^2m_0c^2 + m_0^2c^4 - p_{LAB}^2C^2$$

$$= 2m_0c^2(mc^2 + m_0c^2).$$

For $m >> m_0$,

$$E_{CM}^2 \approx 2m_0c^2mc^2 \approx m_0c^2.E_{LAB}$$

so

$$E_{CM} = \sqrt{2m_0c^2 . E_{LAB}} ,$$

ultimately one must quadruple the lab energy to double the centre of mass energy.

HOW RELATIVITY CONNECTS ELECTRIC AND MAGNETIC FIELDS

A MAGNETIC PUZZLE...

Suppose we have an infinitely long straight wire, having a charge density of electrons of $-\lambda$ coulombs per meter, all moving at speed v to the right (recall typical speeds are centimeters per minute) and a neutralizing fixed background of positive charge, also of course λ coulombs per meter. The current in the wire has magnitude $I = \lambda v$ (and actually is flowing to the left, since the moving electrons carry negative charge).

Suppose also that a positive charge q is outside the wire, a distance r from the axis, and this outside charge is moving at the same exact velocity as the electrons in the wire.

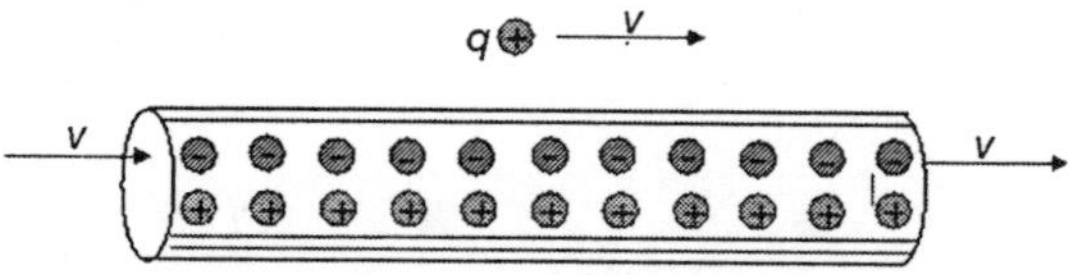

Positive charges (ions) are fixed: negative charages (electrons) drift at speed v. Positive charge outside wire also moves at v.

Note: Positive and negative charges are of course uniformly distributed throughout the wire. They are shown separated here for clarity.

What force does the positive charge q feel?

The wire is electrically neutral, since it contains equal densities of positive and negative charges, both uniformly distributed throughout the wire (the illustration above is of course schematic). So q feels no electrical force.

However, since q is moving, it will feel a magnetic force,

$$\vec{F}_{\text{mag}} = q\vec{v} \times \vec{B}\ .$$

The magnetic field lines are vertical circles centered along the axis of the wire, the field being into the page where the charge is (remember the current is to the left) and of magnitude

$$B = \frac{\mu_0 I}{2\pi r} = \frac{\mu_0 \lambda v}{2\pi r}$$

so the force on the charge is of magnitude

$$F = \frac{q\mu_0 \lambda v^2}{2\pi r}$$

and is directed away from the wire, so the charge will accelerate away from the wire.

Now let us examine the same physical system in the frame of reference in which the charge is initially at rest. In that frame, the electrons are also at rest, but the positive background charge is flowing at v:

The view in the rest frame of the electrons and the external positive charge

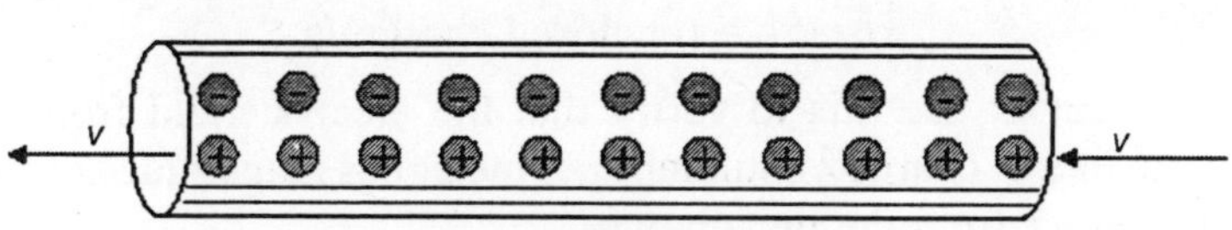

What force does q feel in this frame?

Since q is at rest, it cannot feel a magnetic force: such forces depend linearly on speed!

Yet it looks as if it can't feel an electric force either, because the positive and negative charges in the wire have equal densities, right? This leads to the conclusion that q feels no force at all in this frame, so it won't accelerate away from the wire, as it did in the other frame. This is of course nonsense—so where did we go wrong?

SOLVED ELECTRICALLY!

The mistake was in ignoring the relativistic Fitzgerald-Lorentz contraction, even though the velocities involved are *millimeters per second*! In the frame in which the wire is at rest, the positive and negative charge densities exactly balance, otherwise there will be extra electrostatic fields that the electrons will quickly move to neutralize. However, this *necessarily* means that the densities cannot balance exactly in the frame in which the drifting electrons are at rest.

In the electrons' rest frame, the positive charges, which had density λ coulombs per meter in their rest frame, are moving at speed v, so relativistic contraction will increase their density to:

$$\frac{\lambda}{1-v^2/c^2} \cong \lambda + \frac{\lambda v^2}{2c^2}.$$

On the other hand, in this electron frame the electrons are at rest, so their density is actually less than it was in the frame of the wire, it has decreased by $\lambda v^2/2c^2$.

The net effect is that the wire, electrically neutral in the lab frame, has a positive charge density $\lambda v^2/c^2$ in the frame of the moving electrons (and the outside charge). Recall Gauss's Law relating the total electric field flux through a surface to the enclosed charge,

$$\int \vec{E}\cdot d\vec{A} = (\text{enclosed charge})/\varepsilon_0.$$

Exercise: use this to verify that the electric field from an infinite line of charge λ coulombs per meter has magnitude $\lambda/2\pi r\varepsilon_0$ at distance r from the line charge.

$$F = qE = q\lambda\frac{v^2}{c^2}\frac{1}{2\pi r\varepsilon_0} = \frac{q\lambda v^2}{2\pi r}\frac{1}{\varepsilon_0 c^2} = \frac{q\lambda v^2}{2\pi r}\mu_0$$

From the exercise result, the electrostatic force on the charge q is:

where in the last step we used $1/c_2 = \mu_0\varepsilon_0$.

This purely electrical force is identical in magnitude to the purely magnetic force in the other frame! So observers in the two frames *will* agree on the rate at which the particle accelerates away from the wire, but one will call the accelerating force magnetic, the other electric. We are forced to the conclusion that whether a particular force on an actual particle is magnetic or electric, or some mixture of both, depends on the frame of reference—so the distinction is rather artificial. A Magnetic Puzzl.

Suppose we have an infinitely long straight wire, having a charge density of electrons of $-\lambda$ coulombs per meter, all moving at speed v to the right (recall typical speeds are centimeters per minute) and a neutralizing fixed background of positive charge, also of course λ coulombs per meter. The current in the wire has magnitude $I = \lambda v$ (and actually is flowing to the left, since the moving electrons carry negative charge).

Suppose also that a positive charge q is outside the wire, a distance r from the axis, and this outside charge is moving at the same exact velocity as the electrons in the wire.

What force does the positive charge q feel?

The wire is electrically neutral, since it contains equal densities of positive and negative charges, both uniformly distributed throughout the wire (the illustration above is of course schematic). So q feels no electrical force. However, since q is moving, it will feel a magnetic force,

The magnetic field lines are vertical circles centered along the axis of the wire, the field being into the page where the charge is (remember the current is to the left) and of magnitude so the force on the charge is of magnitude and is directed away from the wire, so the charge will accelerate away from the wire.

Now let us examine the same physical system in the frame of reference in which the charge is initially at rest. In that frame, the electrons are also at rest, but the positive background charge is flowing at v:

What force does q feel in this frame?

Since q is at rest, it cannot feel a magnetic force: such forces depend linearly on speed! Yet it looks as if it can't feel an electric force either, because the positive and negative charges in the wire have equal densities, right? This leads to the conclusion that q feels no force at all in this frame, so it won't accelerate away from the wire, as it did in the other frame. This is of course nonsense—so where did we go wrong?

SOLVED ELECTRICALLY

The mistake was in ignoring the relativistic Fitzgerald-

Lorentz contraction, even though the velocities involved are *millimeters per second*! In the frame in which the wire is at rest, the positive and negative charge densities exactly balance, otherwise there will be extra electrostatic fields that the electrons will quickly move to neutralize. However, this *necessarily* means that the densities cannot balance exactly in the frame in which the drifting electrons are at rest.

In the electrons' rest frame, the positive charges, which had density λ coulombs per meter in their rest frame, are moving at speed v, so relativistic contraction will increase their density to:

On the other hand, in this electron frame the electrons are at rest, so their density is actually less than it was in the frame of the wire, it has decreased by $\lambda v^2/2c^2$.

The net effect is that the wire, electrically neutral in the lab frame, has a positive charge density $\lambda v^2/c^2$ in the frame of the moving electrons (and the outside charge). Recall Gauss's Law relating the total electric field flux through a surface to the enclosed charge,

Exercise: use this to verify that the electric field from an infinite line of charge λ coulombs per meter has magnitude $\lambda/2\pi r\varepsilon_0$ at distance r from the line charge.

From the exercise result, the electrostatic force on the charge q is: where in the last step we used.

This purely electrical force is identical in magnitude to the purely magnetic force in the other frame! So observers in the two frames *will* agree on the rate at which the particle accelerates away from the wire, but one will call the accelerating force magnetic, the other electric. We are forced to the conclusion that whether a particular force on an actual particle is magnetic or electric, or some mixture of both, depends on the frame of reference—so the distinction is rather artificial.

REMARKS ON GENERAL RELATIVITY

EINSTEIN'S PARABLE

In Einstein's little book *Relativity: the Special and the General Theory*, he introduces general relativity with a parable. He imagines going into deep space, far away from gravitational fields, where any body moving at steady speed in a straight line will continue in that state for a very long time. He imagines building a space station out there - in his words, "a spacious chest resembling a room with an observer inside who is equipped with apparatus." Einstein points out that there will be no gravity, the observer will tend to float around inside the room.

But now a rope is attached to a hook in the middle of the lid of this "chest" and an unspecified "being" pulls on the rope with a constant force. The chest and its contents, including the observer, accelerate "upwards" at a constant rate. How does all this look to the man in the room?

He finds himself moving towards what is now the "floor" and needs to use his leg muscles to stand. If he releases anything, it accelerates towards the floor, and in fact all bodies accelerate at the same rate. If he were a normal human being, he would assume the room to be in a gravitational field, and might wonder why the room itself didn't fall. Just then he would discover the hook and rope, and conclude that the room was suspended by the rope.

Einstein asks: should we just smile at this misguided soul? His answer is no - the observer in the chest's point of view is just as valid as an outsider's. In other words, being inside the (from an outside perspective) uniformly accelerating room is physically equivalent to being in a uniform gravitational field. This is the basic postulate of general relativity. Special relativity said that all inertial frames were equivalent. General relativity extends this to accelerating frames, and states their equivalence to frames in which there is a gravitational field. This is called the Equivalence Principle.

The acceleration could also be used to cancel an existing gravitational field—for example, inside a freely falling elevator passengers are weightless, conditions are equivalent to those in the unaccelerated space station in outer space.

It is important to realise that this equivalence between a gravitational field and acceleration is only possible because the gravitational mass is exactly equal to the inertial mass. There is no way to cancel out electric fields, for example, by going to an accelerated frame, since many different charge to mass ratios are possible.

As physics has developed, the concept of fields has been very valuable in understanding how bodies interact with each other. We visualize the electric field lines coming out from a charge, and know that something is there in the space around the charge which exerts a force on another charge coming into the neighborhood. We can even compute the energy density stored in the electric field, locally proportional to the square of the electric field intensity. It is tempting to think that the gravitational field is quite similar—after all, it's another inverse square field. Evidently, though, this is not the case.

If by going to an accelerated frame the gravitational field can be made to vanish, at least locally, it cannot be that it stores energy in a simply defined local way like the electric field.

We should emphasize that going to an accelerating frame can only cancel a *constant* gravitational field, of course, so there is no accelerating frame in which the whole gravitational field of, say, a massive body is zero, since the field necessarily points in different directions in different regions of the space surrounding the body.

SOME CONSEQUENCES OF THE EQUIVALENCE PRINCIPLE

Consider a freely falling elevator near the surface of the earth, and suppose a laser fixed in one wall of the elevator sends a pulse of light horizontally across to the corresponding point on the opposite wall of the elevator. Inside the elevator, where there are no fields present, the environment is that of an inertial frame, and the light will certainly be observed to proceed directly across the elevator. Imagine now that the elevator has windows, and an outsider at rest relative to the earth observes the light.

As the light crosses the elevator, the elevator is of course accelerating downwards at g, so since the flash of light will hit the opposite elevator wall at precisely the height relative to the elevator at which it began, the outside observer will conclude that the flash of light also accelerates downwards at g. In fact, the light could have been emitted at the instant the elevator was released from rest, so we must conclude that light falls in an initially parabolic path in a constant gravitational field.

Of course, the light is traveling very fast, so the curvature of the path is small! Nevertheless, *the Equivalence Principle forces us to the conclusion that the path of a light beam is bent by a gravitational field.*

The curvature of the path of light in a gravitational field was first detected in 1919, by observing stars very near to the sun during a solar eclipse. The deflection for stars observed very close to the sun was 1.7 seconds of arc, which meant measuring image positions on a photograph to an accuracy of hundredths of a millimeter, quite an achievement at the time.

One might conclude from the brief discussion above that a light beam in a gravitational field follows the same path a Newtonian particle would if it moved at the speed of light. This is true in the limit of small deviations from a straight line in a constant field, but is not true even for small deviations for a spatially varying field, such as the field near the sun the starlight travels through in the eclipse experiment mentioned above.

We could try to construct the path by having the light pass through a series of freely falling elevators, all falling towards the centre of the sun, but then the elevators are accelerating relative to each other (since they are all falling along *radii*), and matching up the path of the light beam through the series is tricky. If it is done correctly (as Einstein did) it turns out that the angle the light beam is bent through is twice that predicted by a naove Newtonian theory.

What happens if we shine the pulse of light vertically *down* inside a freely falling elevator, from a laser in the centre of the ceiling to a point in the centre of the floor? Let us suppose the flash of light leaves the ceiling at the instant the elevator is released into free fall. If the elevator has height h, it takes time h/c to reach the floor. This means the floor is moving downwards at speed gh/c when the light hits.

Question: Will an observer on the floor of the elevator see the light as Doppler shifted?

The answer has to be no, because inside the elevator, by the Equivalence Principle, conditions are identical to those in an inertial frame with no fields present. There is nothing to change the frequency of the light. This implies, however, that to an outside observer, stationary in the earth's gravitational field, the frequency of the light *will* change.

This is because he will agree with the elevator observer on what was the initial frequency f of the light as it left the laser in the ceiling (the elevator was at rest relative to the earth at that moment) so if the elevator operator maintains the light had the same frequency f as it hit the elevator floor, which is moving at gh/c relative to the earth at that instant, the earth observer will say the light has frequency $f(1 + v/c) = f(1+gh/c^2)$, using the Doppler formula for very low speeds.

We conclude from this that light shining downwards in a gravitational field is shifted to a higher frequency. Putting the laser in the elevator floor, it is clear that light shining upwards in a gravitational field is red-shifted to lower frequency. Einstein suggested that this prediction could be checked by looking at characteristic spectral lines of atoms near the surfaces of very dense stars, which should be red-shifted compared with the same atoms observed on earth, and this was confirmed. This has since been observed much more accurately.

An amusing consequence, since the atomic oscillations which emit the radiation are after all just accurate clocks, is that *time passes at different rates at different altitudes*. The US atomic standard clock, kept at 5400 feet in Boulder, gains 5 microseconds per year over an identical clock almost at sea level in the Royal Observatory at Greenwich, England. Both clocks are accurate to one microsecond per year. This means you would age more slowly if you lived on the surface of a planet with a large gravitational field. Of course, it might not be very comfortable.

GENERAL RELATIVITY AND THE GLOBAL POSITIONING SYSTEM

Despite what you might suspect, the fact that time passes at different rates at different altitudes has significant practical consequences. An important *everyday* application of general relativity is the Global Positioning System. A GPS unit finds out where it is by detecting signals sent from orbiting satellites at precisely timed intervals. If all the satellites emit signals simultaneously, and the GPS unit detects signals from four different satellites, there will be three relative time delays between the signals it detects. The signals themselves are encoded to give the GPS unit the precise position of the satellite they came from at the time of transmission. With this information, the GPS unit can use the speed of light to translate the detected time delays into distances, and therefore compute its own position on earth by triangulation.

But this has to be done very precisely! Bearing in mind that the speed of light is about one foot per nanosecond, an error of 100 nanoseconds or so could, for example, put an airplane off the runway in a blind landing. This means the clocks in the satellites timing when the signals are sent out must certainly be accurate to 100 nanoseconds a day.

That is one part in 10^{12}. It is easy to check that both the special relativistic time dilation correction from the speed of the satellite, and the general relativistic gravitational potential correction are much greater than that, so the clocks in the satellites must be corrected appropriately. (The satellites go around the earth once every twelve hours, which puts them at a distance of about four earth radii. The calculations of time dilation from the speed of the satellite, and the clock rate change from the gravitational potential, are left as exercises for the student

In fact, Ashby reports that when the first Cesium clock was put in orbit in 1977, those involved were sufficiently skeptical of general relativity that the clock was not corrected for the gravitational redshift effect. But—just in case Einstein turned out to be right—the satellite was equipped with a synthesizer that could be switched on if necessary to add the appropriate relativistic corrections.

After letting the clock run for three weeks with the synthesizer turned off, it was found to differ from an identical clock at ground level by precisely the amount predicted by special plus general relativity, limited only by the accuracy of the clock. This simple experiment verified the predicted gravitational redshift to about one percent accuracy! The synthesizer was turned on and left on.

Chapter 10

Particle Physics

ELEMENTARY PARTICLES

The first elementary particle, the electron, was discovered by Thompson. But elementary particle physics as a separate area of physics was really started in the 1940's and 1950's, some time after the discovery of proton, neutron, and positron, that are all elementary particles (at least they were regarded as such until about 1970). There were two important reasons for the blossoming of elementary physics, one experimental and one theoretical.

The experimental impetus was given by the observation that very high energy particles, first observed in cosmic radiation, are able to initiate reactions, similar to nuclear interactions, when collide with the nuclei of O, N, H,... of the atmosphere. It was observed, using photographic emulsions flown at high altitude, that when particles of the cosmic radiation (mostly protons) collide with nuclei they produce a large number of new particles, at very narrow forward angles, sometimes a far larger number than the number of nucleons in the target nucleus.

The large cross section for these collisons and the shear number of particles produced were a proof that the interaction between the nucleus and the incoming cosmic ray particle is very strong. Thus, it was immediately thought that most of the particles produced were not nucleons, but rather the quanta of the strong force.

The existence of these quanta, called p-mesons, was predicted on purely theoretical grounds by Hideki Yukawa in the 1930's. It was soon proved that indeed most of the particles produced of these high energy collisions are p-mesons. Their approximate mass was also predicted successfully based on the structure of Yukawa interactions,

$$V_{Yukawa} = -\,g^2 e^{-\,m\,r\,c/\,hbar/}\ r = -\,g^2 e^{-\,r/\,R}$$

where m is the mass of the particle communicating the interaction. Note that electromagnetic interactions are similar if we take $g^2 \rightarrow k\,e^2$ and m=0 (the mass of the photon is 0!). By estimating the range of nuclear interactions, which is equal to R = hbar/(m c) = hbar c/ (m c^2) one is able to come up with a number

$$mc^2 = \text{hbar c/ R} = 197.3\ \text{MeV f/}\ 1.2\ \text{f} = 164\ \text{MeV}.$$

The true mass of p-mesons is 138 MeV, remarkably close to this crude estimate. The investigation of interactions of nucleons and p-mesons lead to the development of elementary particle physics. The other ingredient in the emergence of elementary particle physics was the success of quantum field theory that became and still is the only successful framework of describing the interactions (creation, annihilation, and scattering) of elementary particles. Quantum field theory was started by Dirac around 1930 from the marriage of relativity and quantum mechanics.

Though the theoretical framework was understood quanum field theory was useless for most practical reasons because calculations, beyond a trivial first approximation cannot be made any sense, they were plagued by divergences (infinities) that could not be eliminated. It took almost two decades until Tomonaga, Schwinger, and Feynman could device a self-consistent method of eliminating these divergences. This method is called renormalization.

The basic idea behind renormalization can be understood easiest in is the framework of Quantum electrodynamics, the quantum field theory describing the interaction of photons with charged particles, such as electrons.

Quantum field theory implies that the vacuum is not really empty, but filled with "virtual" particles that are keep being produced and annihilated. Conservation of energy momentum is satisfied because the mass-energy of these virtual particles does not equal to the mass-energies observed for physical particles traveling in the vacuum. Virtual particles live for a very short time, before annihilating, thus the uncertainty relation between time and energy requires that their mass-energy is undetermined to a large degree. The presence of these virtual particles on real electrons is profound and twofold. The mass-energy of the real electron is changed substantially by interactions with the bath of virtual particles around it. Note that even classically the self energy of a pointlike electron is infinite. It is also infinite, due to interactions with virtual particles, in quantum field theory.

These interactions will give an infinite contribution to the mass energy of electrons. It turns out, however that if the mass-energy of the electron (the so-called bare mass-energy) without the interaction with virtual particles is chosen to be infinitely large and negative then the sum of contributions can be made finite and one can end up with a finite mass in a consistent series of approximations of arbitrary precision.

Similarly, the other parameter describing electrons, their charge, can also be treated in a similar manner. Note that the presence of the negatively charged electron polarizes virtual particles, the total charge of which must be equal to zero.

Thus, the electron sits in a polarized dielectric medium. Then, observed from practically infinite distance (that is what experimental observations practically do) the charge of the electron seems considerable smaller due to screening. In fact if, starting form infinity, one would be able to go closer and closer to the electron, one would observe a larger and larger negative charge.

One can show that if one wishes to keep the observed charge (the one observed from infinite distance) finite then the charge one would observe at zero distance from the electron (the so-called bare charge), when all the layers of screening are peeled away, would have to be infinitely large negative.

Conversely, chosing the "bare charge" to be infinite one can build up a scheme of systematic calculations that can in principle performed to arbitrary precision. Calculations performed following the method of Tomonaga, Schwinger, Feynman and Dyson had a tremendous success in explaining experiments of ultimate precision, such as the anomalous magnetic moment of electrons and the Lamb shift.

CONSERVATION LAWS

After the discovery of pions, other elementary particles were discovered. Pions, or p-mesons, are unstable. That is why we do not see them under normal circumstances. The lifetime of π^+ (or π^-) mesons is not very short, by elementary particle standards, $\tau = 2.6 \times 10^{-8}$ s. That means, that relativistic π^+ mesons can fly to a distance of hundreds of meters in the laboratory, if we take into the effect of time dilation. In fact, these days pion beams are produced from targets at accelerators provide an everyday experimental tool.

π^+– mesons decay into m^+ mesons and neutrinos, $\pi^+ \rightarrow \mu^+ + \nu$. μ^+ are of course lighter than π^+, but they are not strongly interacting, only electromagnetically (they are charged!) therefore they penetrate matter readily, they can traverse thick steel slabs without interactions. This property is used for their identification. In general, when elementary particles are studies it is extremely important to identify as many of the produced particles as possible. This helps understanding the basic laws of nature.

The antiparticles of π^+ mesons are π^--mesons, they decay into $\pi^- \rightarrow \mu^- + \nu^c_\mu$ (where this latter is a muon antineutrino, as distinguished form a muon neutrino, or an electron neutrino). Muons, or m-mesons, behave much the same way as electrons (or positrons). They only have electromagnetic and weak (like beta decay) interactions.

Only their mass (106 MeV/c^2) distinguishes them from electrons. It is still not understood why nature needs to duplicate electrons in this fashion. Along with ν_μ they carry an additive quantum number, muon number that has been observed to conserve in all interactions. The electron number carried by electrons and electron neutrinos is also conserved (positrons and antineutrinos carry electron number +1). The muons decay as

$$\mu^+ \rightarrow e^+ + \nu_\mu + \nu_e$$

in about 10^{-6}s. Muons and electrons, with their respective neutrinos are called leptons.

After the discovery of pions by Lattes, Occhialini, and Powell other particles were discovered in rapid pace. First, it was observed that near the collisions particle pairs, in the form of a narrow V diracted toward the collision centre, seemed to emerge from nothing. This was immediately interpreted as the decay of a hitherto unknown neutral particle into two positive particles. These

V-events were later identified as the decay of K-mesons and hyperons. While K-mesons (m= 494 MeV/c^2) are similar to p - mesons, they do not carry a nucleon number, hyperons are different (The lightest hyperon is the L^0 hyperon of m=1116 MeV/c^2). This is shown by their decays:

$$K^+ \rightarrow \pi^+ + \pi^0,$$

$$\Lambda^0 \rightarrow p + \pi^- \text{ or } n + \pi^0$$

As if a hyperon would have nucleon inside, as this is already indicated by its large mass. Thus, we need to expand the notion of nucleon number to baryon number (baryos=heavy in greek). Nucleons and hyperons are baryons and it is not the nucleon number, A, that is conserved but the baryon number (B). This is clearly shown in production processes. Hyperons, just like nucleons, have s=1/2, while pions and kaons are s=0 bosons. As physicists learned how to build accelerators they succeeded in creating beams of various particles (by separating the products of p-p collisions by poweful magnets). Then they observed reactions as (that is a strong interaction with large cross section as 30 mb)

$$\pi^+ + n \rightarrow \Lambda^0 + K^+.$$

Obviously in the above process baryon number is conserved but nucleon number is not. It was remarkable that reactions like were never observed.

$$\pi^+ + n \rightarrow \Lambda^0 + \pi^+$$

This required the introduction of a new quantum number, strangeness, by Gell-Mann and Pais, such that S(Λ^0) = –1, S(K^+) =1. At the same time in decay processes such as K-decay or hyperon decay strangeness was not conserved. This was the first case that some quantum numbers were conserved in strong interactions and electromagnetic interactions, but not in weak interactions. These are very well distinguishable, the characteristic reaction time of one is 10^{-23} s while for the other 10^{-10} s.

Other hyperons were also discovered soon, Σ^+, Σ^- and S^0 hyperons that all had S = –1 (of course their antiparticles have S = + 1), the Ξ^- and Ξ^0 hyperons that have S = –2. All these hyperon have B = 1, just like the proton and the neutron, say B(Ξ^0) =1. Thus Ξ^0 is produced in a reaction like

$$K^- + p \rightarrow \Xi^- + K^+$$

Note that in this reaction both B ans S are conserved.

Hyperons are all unstable, after a characteristic time of about 10^{-10} s they decay into nucleons and mesons or leptons. Example:

$$\Xi^- \rightarrow \Lambda^0 + \pi^- \text{ or } \Lambda^0 + \mu^- + \nu_\mu.$$

In decay processes (weak interactions) the strangeness can only change by only one unit. Note the second of these decay processes, it is just like beta decay. Even more similar to beta decay are two modes of L^0 decay (they happen very rarely)

$$\Lambda^0 \rightarrow p + e^- + \nu_e$$

$$\Lambda^0 \rightarrow p + \mu^- + \nu_m$$

There are other additive quantum numbers, called flavors that are similar to strangeness and are also conserved in strong interactions.

Another important quantum number conserved in strong intreractions is isotopic spin. It has an important role in nuclear physics, as well. It was noticed very early that just like nucleons (p,n) all strongly interactive particles come in multiplets having equal, or almost equal mass.

Examples are (π^-, π^0, π^+), (K^+, K^0), Λ^0 (alone), (Σ^+, Σ^0, Σ^-), (Ξ^-, Ξ^0),... These multiplets (singulet, doublet, triplet,...) have the important property that their members have the same spin, baryon number, strangeness, and approximate mass. They have identical strong interactions.

Apart from their small mass differences they are distinguished by their charge (and consequently electromagnetic interactions). In this respect they are very similar to a system with a fixed spin that can have states with different spin projections. E.g. we have a s=1 system that can have $s_z = -1,0$, and 1. Pions, having three charge states have I = 1, so they can have $I_3 = -1,0$ and 1, corresponding to the states of π^-, π^0, and π^+. Experiments showed that isotopic spin is conserved in all strong interactions. E.g. in the collision of a π^+ and a p the total I can either be I=1/2 or I=3/2. Then the final state products must also have either I=1/2 or I=3/2.

Nucleons have isotopic spin I=1/2, since they have 2I+1 = 2 charge states, p and n. Then isotopic spin is very relevant for nuclear physics, because nuclear physics is mostly a study of strong interactions. Then combining two nucleons can lead to I=0, or 1. The only two nucleon state is d. Then d has I=0. Nuclear forces depend strongly on isospin there are no nn or pp bound states. One could explain away the state pp by invoking the Coulomb repulsion, but there is no other good reason to forbid the nn state.

Since the introduction of isotopic spin is the recognition that p and n are just two charge states of the same particle, the Pauli exclusion principle should also be applied to them. Investigate the d from the point of view of antisymmetry of the two nucleons. The p and the n, being in I=0 state, is antisymmatric for the exchange of the p and n. (remember that s=1 is a symmetric state and s=0 is an antisymetric state). Then their wave function must be symmetric in spin, because the overall wavefunction must be antisymmetric for fermions. If the spin wavefunction is symmetric then the spin must be s=1. Indeed, the s(d)=1.

If one goes to heavier nuclei, then it is obvious that the pair ($^3_2$He, $^3_1$H) forms an isodoublet. Their collision with d can only produce an I=1/2 state in strong interactions. So

$$d + {}^3_2\text{He} \rightarrow {}^4_2\text{He} + p$$

happens because I($^4_2$He) =0. Electromagnetic interactions violate I conservation, though I_3 is still conserved.

PARTICLE INTERACTIONS, RESONANCES, AND QUARKS

After the discovery of mesons and hyperons experimentalists found a large number of excited states of these particles as well. The typical example of these resonances is the Δ particle that has four charge states (isotopic spin, I=3/2) Δ^{++}, Δ^{+}, Δ^{0}, and Δ^{-} This resonance was found by Fermi in an experiment in which he bombarded protons (hydrogen) by a newly created p-meson been at the Bevatron accelerator in Berkeley. The reaction he observed was

$$\pi^- + \pi \rightarrow \Delta^0 \rightarrow \pi^- + p$$

Now the intermediate state, Δ^0 is of extremely short lifetime. In fact, t ~ 10^{-23} s. Direct observation of such a particle is impossible. The uncertainty relation between time and energy, however, implies that the energy of the Δ is uncertain to about 100 MeV. The Δ^-p cross section increases when the total energy-momentum squared $E^2 = (E_p + E_\pi)^2 - c^2 (p_p + p_\pi)^2 = m_\Delta{}^2 c^4$ This is so because there is a new channel throgh which the scattering can happen. Due to the uncertainty of m_Δ, $\delta\, m_\Delta \sim$ hbar/ ($c^2\, \tau$), the maximum in the p^- p cross section is not sharp, but smeared over a region of about 100Mev in the primary energy of π^-. Thus, if we plot the cross section as a function of the energy of the pion (performing the experiment at a series of different energies) then we will see a large bump at $E^2 = m_\Delta^{\,2}\, c^4$. The width of the bump provides the lifetime of the Δ particle. This particle is a baryon if you require baryon number conservation. It also turns out that the spin of this particle is s =3/2. The mass of Δ is m_Δ=1232 MeV/ c^2

As time went by many more particle resonances were discovered. These resonances can be regarded as new particles on their own right. They just happen to have large enough masses so that conservation laws allow them to decay via strong interactions into strongly interacting particles. That is the reason why their lifetime are so short. As an example the mass of Δ happens to satisfy $m_\Delta > m_\pi + m_p$ so Δ is allowed to decay into π + p. After literally hundeds of resonances with exactly the same quantum numbers as some other particles were discovered the situation became untenable.

This looked more or more as the a system of states with the excitations of the same system, like an atom or a nucleus. The existence of such excitations is a sure sign of composite nature. Clearly atoms and nuclei are composite. This gave the idea to Nambu and Gell-Mann to postulate the existence of constituents for strongly interacting particles: quarks.

The most economic choice (due to Gell-Mann) turned out to be very strange: All normal and strange strongly interacting particles in every charge mode can be constructed out of three quarks, and their antiparticles (antiquarks). The three quarks are u (up), d (down), and s (strange) quarks. Their charge are not integer times e.

quark	*charge*	*spin*	*Baryon number*	*Strangeness*	*Isotopic spin*	I_3
u	2/3 e	1/2	1/3	0	1/2	1/2
d	–1/3 e	1/2	1/3	0	1/2	–1/2
s	– 1/3 e	1/2	1/3	–1	0	0

If one uses these quantum numbers then all states known at

that time could be built from the combination of these quarks. Examples: $\pi^- = du^c$ (where I denote antiparticles by superscript c. Ususal notation is a bar above the u, or d, or s), p = uud, n = udd, Δ^{++} = uuu, K^- = su^c, Λ^0 = sdu, etc. Note that one gets the right spins and isotopic spins as well.

Baryons, made of three quarks, must be fermions. In the ground state (zero orbital momentum) they must have spins of s = 1/2 or s = 3/2. Indeed the nucleons have s=1/2 and the D baryons have s = 3/2. Since both of these states are made of three I=1/2 quarks (u and d) their isospin must be 1/2 or 3/2 as well. Of course, we know that I(N) = 1/2 and I(Δ) = 3/2, because they have 2 and four charge states respectively.

Anyway, the existence of Δ^{++} demands I = 3/2, because I_3 is an additive quantum number and π^{++} = uuu. Each u has I_3 = 1/2 so the total I_3=3/2. But in that case I cannot be smaller than I=3/2. Similarly, Σ^+ = uus, so, since I(s) = 0, it must have I = 1. Indeed it has three charge states. Finally, the starngeness of Ξ S(Ξ) = -2 so it must contain 2 strange quarks. Concequently, Ξ^- = ssd, Ξ^0 = ssu, they have isospin I(Ξ) =1/2, because they contain one I = 1/2 quark (u or d) and two I = 0 quarks (s). The above list conatins all ground states of baryons constructed from the three quarks u, d, and s, except sss. This state was predicted by Gell-Mann and subsequently discovered and was called Ω^-.

After the identification of the ground state baryons the previously discovered resonances of higher mass and frequently higher spin were also identified as excited states of these meson and baryon states, with nonzero orbital angular momenta between quarks. Subsequently, heavier and heavier states were investigated, mostly in $e^+ - e^-$ colliders. On theoretical grounds, the existence of another quark, the charmed quark was predicted. c- anti c bound states were found as tremendously high peaks in the $e^+ - e^-$ cross sections near total energy (in the CMS) 3 GeV. These peaks were very narrow O(100 keV), which cannot be explained any other way but by the interpretation that such a new state, so called charmonium was produced (Richter and Ting got the Nobel prize for the discovery).

The fact is that the decay is from strongly interacting particles to strongly interacting particles should be much faster (i.e. have larger width). QCD, the theory of strong interactions. While the u d and s quarks are light, M(c) ~ 1.5 GeV/c^2

Subsequently, and unexpectedly, two more quarks, b (bottom) and t (top) (other two Nobel prizes) were discovered in similar typ of experiments. Their masses are large M(b) = 5 GeV/c^2 and M(t) = 170 GeV/c^2 (as heavy as a A=180 nucleus). Furthermore, a third lepton family was also discovered (also Nobel prize winner), with the t (positive and negative) lepton and corresponding neutrino.

There is fairly strong (astrophysical, and accelerator physics) evidence that there are no more quarks or leptons. The known leptons and quarks can be grouped into three families. First family: d, u, e, ν_e, Second family: s, c, m, ν_μ and the third family: b, t, t, ν_τ. These are listed in order of increasing mass (except possible neutrinos, whose mass is consistent with zero). All ordinary matter, including radioactive decay products are made of the members of the first family. As we seę it, the world would not be substantially different if the second and third families would not exist.

It is still a mistery why the first family is duplicated twice. The additive quantum numbers distinguishing the six quarks, charge, strangeness, char, and the bottom and top quantum numbers are called flavors. They are all conserved in strong and electromagnetic interactions, but with the exception of charge they all can be violated in weak interactions.

THE STANDARD MODEL

There is a substantial problem with the quark assignments of baryons. Take for example the Ω^- hyperon. Its assignment is sss and its spin is s(Ω^-) = 3/2 This spin state is completely symmetric for the exchange of the s quarks (all the three spins are parallel). Consequently the wave function of the Ω^- hyperon is a completely symmetric combination of the wavefunctions of the three s quarks. Similarly the Δ baryons have isospin I = 3/2 and spin I = 3/2, both of these states are completely symmetric, i.e the three quark wavefunctions are completely symmetric.

In fact, assuming that the three quark wavefunctions are completely symmetric one can prove that the baryon ground states formed from threee u and d quarks are exactly the observed isospin

I=3/2 and spin I=3/2 state (D) and an I=1/2 and spin I=1/2 state (the nucleons). This is very troublesome because it requires the symmetrizations of wavefuncions in fermions, while according to the Pauli exclusion principle the wavefunctions should be completely antisymmetric.

Now the symmetry properties of wavefunctions and their relation to the bosonic and fermionic nature of particles is a very basic law of physics and should not be violated. This gave a lot of headache to theorists, who finally came up with the right solution. The only way to solve the situation is that the the three s quarks that make up Ω^{-}, or the three u quarks that make up Δ^{++} differ in an additional, hitherto unknown, quantum number, so they can be antisymmetrized in that quantum number. Gell-Mann called this quantum number colour (he also named quarks). Thus there are three colors: say red, blue, and yellow and then the wavefunctions of the three quarks inside the baryon can be completely antisymmetrized in colour, requiring that the three quarks are completely symmetrized in the remainging quantum numbers, such as spin and isospin.

One can show using group theory, that such an antisymmetric combination of colour states is "colorless" does not have any colour quantum numbers. In a similar way mesons are constructed by equal combination red-antired, blue-antiblu, and yellow-antiyellow quarks, that is also a colorless state. Furthermore, one can see that the baryon and meson states of this kinds are the only colorless states one can construct from three quarks and a quark antiquark pair respectively.

Since free quarks have never been seen in nature the challange is to construct a theory of quarks that allows the existence of colorless physical states constructed from quarks only. Such a theory is quantum chromodynamics (QCD).

Quantum chromodynamics is built based on the analogy of Quantum electrodynamics, the theory of interaction of photons with charged particles. The strength of electromagnetic interactions is given by e. In field theory, so a similar "strong coupling constant," g, is defined.

Since g is much larger than e, strong interactions are stronger than electromagnetic interactions. The role of photons is to generate interactions among charged particles. There should be particles call gluons, that create interactions among particles that have the colour quantum number. The difference is that while there is only one kind of a conserved charge in electromagnetism, in QCD there are three colors to play a similar role. Accordingly, there are much more kinds of gluons (eight) that communicate interactions among objects having colour quantum numbers. While in Quantum electrodynamics an photon can turn into an electron positron pair, gluons can turn into quark-antiquark pairs.

So there is one gluon that can turn into a blue-antired quark pair, etc. The great difference is that while photons themselves are not charged (the total charge of the electron positron pair is zero), gluons themselves carry colour. Since photons interact with charged objects only, tnot being charged, they do not directly interact with each other. At the same time gluons, carrying colour charges do have direct interactions with each other.

The fact that gluons interact directly with each other has far reaching consequences in strong interactions. Remember, that the vacuum screening of charges gets stronger and stronger as one goes farther and farther away from the electron. The consequence is that the effective charge is decreasing with distance. Alternatively, using the uncertainty relation that says that large distance corresponds to low energy and short distance corresponds to high energy. The one can show that the effective charge has an energy dependence, i.e. at low energies a lower effective charge can be used in calculations (of cross sections, decay rates, etc.), but that increases at high energies the effective charge is larger. Now in QCD, due to the self interaction of gluons the screening situation is entirely different.

The gluons that pop out from the vacuum generate antisceening, because the like colour charges and anticharges repel each other. The end result is that at large distances (or at low momentum) the effective colour charge (g) increases, while at short distances (high momentum) the effective colour charge, g decreases. One can also show that in the limit of of infinite momentum (or energy) the effective charge vanishes. This property is called asymptotic freedom.

If the effective charge is very small at high energies then the basic calculational method of field theory of expansion in a power series of charge works very well. Indeed very highenergy experiments confirm this they agree completely with the predictions of QCD, though the precision is far less than the experimental agreement with predictions of QED.

At low energies the effective charge is large and the calculations break down. The only results at low energies are from crude numerical symulations that confirm the property of confinement: all states that carry a colour quantum number (single quark, single gluon, diquark) are non existent in the free form, quarks and gluons are permanently confined inside mesons and baryons.

The picture of strongly interactive particles is then the following: If we bombard a proton with very high energy probes then it behaves as a collection of three free quarks. This can be clearly seen in experiements. The very high energy probes (think of the uncertainty relation again) probe the proton at very short distances. So if the quarks are close they interact very little. Then when we try to separate them their interaction is getting stronger. When they are at a larger distance, the attractive energy between themm increases proportionally to the distance. To separate them to infinite distance we must supply infinite energy. Thus, they are permanently confined inside mesons and baryons.

If quarks are never seen in nature in the free form the question can be raised whether they exist at all or they are some kind of a mathematical concept only, a mnemonic that tells you how to construct elementary particles. There is extensive evidence for the existence of quarks. One is the bombardment of protons by very high energy electrons. Electrons do not participate in strong interactions, only in electromagnetic and weak interactions.

Negelecting weak interactions, the scattering is purely electromagnetic. At very high energy the proton looks like a collection of free quarks. Scattering on free quarks can be easily calculated (something like Rutherford scattering) and the calculated cross sections agree precisely with the experimentally observed ones. More direct and specttacular evidence comes from high energy $e^+ - e^-$ colliders. When these particles collide in th eCMS (which is the laboratory) they create a virtual photon at rest in the laboratory, but having an enormous mass. The photon (if it is not polarized) loses the directionality of the original e^+ and e^- comlpetely. Then it is spectacular to see that the large number of strongly interacting particles produced and observed in the final state come out in two narrowly focused jets in oppositie direction.

Without quarks one would expect a more or less isotropic distribution for these produced particles. Now how do quarks explain the existence of jets? At very high energy quarks interact with the photon (they are charged!) but their strong interactions are "weak" so the only thing that happens is that the virtual photon can decay into quark-antiquark pairs, which, due to energy-momentum conservation fly out in opposite directions with very high momentum.

As long as the quarks are not very far from each other they continue to fly and have still weak interactions. When the distance between them is increasing the potential energy also increases and try to stop them. Instead of stopping them virtual quark -antiquark pairs and gluons are produced from the vacuum and join the initial quarks in a way that they become colorless object that are now free to go out to infinite distance. Since this happens after the original quarks were produced their directionality is conserved and two jets are produced. More precise calculations in QCD confirm this scenario. There is no other known theory that could produce jets as QCD does.

NUCLEAR PHYSICS

The relevant length scale for measuring nuclear size is the femtometer 1 fm = 10^{-15} meters. Physicists usually call this length one *fermi*. Nuclei vary from about one to a few fermis in radius. Recall that the Bohr radius of Hydrogen is of order 10^{-10} meters, so the nucleus is far smaller than the atom. Nuclear size was first measured by Rutherford, by noting how close α-particles came to the nucleus before the scattering ceased to be pure Coulomb repulsion (at which point they were actually hitting the nuclear surface). Despite its small size, the nucleus has about 99.97% of the mass of the atom (a bit less for hydrogen).

WHAT IS THE NUCLEUS MADE OF?

The simplest nucleus, that of hydrogen, is a single proton: an elementary particle of mass about 940 MeV, carrying positive charge exactly opposite to the electron's charge, having a spin of one half and being a fermion (so no two protons can be in the same quantum state).

The next simplest nucleus, called the *deuteron*, is a bound state of a proton and a neutron. The neutron, like the proton, is a spin one-half fermion, but it has no electric charge, and is slightly heavier (by 1.3 MeV) than the proton. The binding energy of the deuteron (analogous to the 13.6 eV for the Hydrogen atom) is 2.2MeV. A photon of this energy could "ionize" the deuteron into a separated proton and neutron. However, it is not necessary to actually do this experiment to establish how tightly the deuteron is bound. One need only weigh the deuteron accurately. It has a mass of 1875.61MeV. The proton has a mass of 938.27MeV, the neutron 939.56MeV, so together (but some distance apart!) they have a mass of 1877.93MeV, 2.2MeV *more* than the deuteron. Thus, when a proton and a neutron come together to form a deuteron, they must unload 2.2MeV of energy, which they do by emitting a photon (called a γ-ray at these energies).

Both protons and neutrons, being fermions, obey the exclusion principle, two protons with spin up cannot be in the same state, although two with opposite spin directions could, and a proton and a neutron can occupy the same spot at the same time!

Protons and neutrons are referred to as *nucleons*. The total number of nucleons in a nucleus is usually denoted by A, where $A = Z + N$, Z protons and N neutrons. The chemical properties of an atom are determined by the number of electrons, the same as the number of protons Z. This is called the *atomic number*. Nuclei can have the same atomic number, but different numbers of neutrons. These nuclei are called *isotopes*, the Greek for "same place", since they are in the same place in the periodic table.

These nucleons attract each other with a short range but very strong force, called the *nuclear force*. The situation here is different from that for electrons in the atom, where the strong central force tends to dominate. In the nucleus the nucleons are attracted mainly by their immediate neighbors. Nevertheless, it is a useful beginning to think of this attractive force as being a potential well, as seen by an individual nucleon, and think in terms of the nuclei as filling the lowest available quantum states in this well, just as we did for electrons in the atom. We find, for example, that the Helium nucleus, $2p+2n$, (a.k.a. the α-particle) is tightly bound—the four nucleons can all occupy the lowest state in the well. However, some larger nuclei, like C, O, Fe are actually a little more tightly bound even than He (about 8.5 MeV per nucleon as opposed to about 7.5 for He) because each nucleon is attracted to its close partners, and there are more close partners in these larger nuclei. It has also been argued that some of these higher nuclei strongly resemble bound states of α-particles.

The total binding energy (usually expressed per nucleon) of any nucleus is easy to find—just as for the deuteron above, the mass of the nucleus is found accurately, and subtracted from the sum of the masses of the separate nucleons.

COULOMB REPULSION AND Á-DECAY

As one goes to really large nuclei, the binding per nucleon actually decreases. This is a consequence of the electrostatic repulsion between the protons. The essential point is that the Coulomb repulsion is *long range*, but the nuclear attraction is short range. Considering a proton at the surface of a nucleus, it is bound to the nucleus by the attractive forces of its nearest neighbors, say five or six of them, and this number is little affected as one goes up the periodic table from, say, Fe to U. On the other hand, the proton on the surface is being pushed away by the electrostatic repulsion from the other protons, and this increases substantially as the number of *other* protons goes from 25 in iron to 91 in uranium, even though they may be a little further away on average. In fact, this is why there is a limit on the size of atoms.

A large nucleus does not, however, tend to spit out protons—there's a much more economical way to unload charge. It ejects an α-particle. The binding energy of the nucleons in an α is almost exactly the same as it is in a U nucleus, so little nuclear binding has to be traded for lowering the electrostatic potential energy.

Thus the most common form of Uranium, U^{238} (A=238, Z=92) goes to Thorium (A=234, Z=90) by α-decay. The surprising thing is, it takes a time comparable to the age of the universe to do so! If it's energetically desirable, why does it take so long?

The answer is—there's a barrier in the way. It is a good model to imagine the U to have the a particle inside it, confined to a deep square well potential representing the attractive nuclear force. However, the α-particle wavefunction will penetrate some distance into the wall of this finite square well potential, where the α will be away from the nuclear force, but not away from the long-range electrostatic repulsion of the other protons in the nucleus. The picture is, then, that outside the well, the ground is not flat it's like a well at the top of a hill.

The implication of this is that if the energy level of the a inside the well is higher than the flat land surrounding the hill, the α-wavefunction after dying away underground as it tunnels out from the well will re-emerge at some distance, oscillating. In other words, there is some finite probability of finding the α shooting away from the nucleus some distance away.

The actual probability of this happening depends on how much the wavefunction decayed as it tunneled underground away from the well, which depended on how far its energy was below the ground. For U^{238}, this decay on tunneling away from the well is so rapid that the α takes on average billions of years to escape. However, many different nuclei undergo α decay, the quickest in millionths of a second. The energies of the emitted α's can be measured, and confirm this explanation of quantum mechanical tunneling, first put forward in 1928. The quickest decays correspond to the most energetic α's, just as the tunneling picture would suggest.

β-DECAY

These heavier nuclei have more neutrons than protons, because the Coulomb repulsion makes it harder to bind protons. However, a price is paid—thinking of a shell-type model, the extra neutrons go into higher energy shells. Thus for nuclei of a given size, there is an optimum ratio of neutrons to protons, close to one for small nuclei where the Coulomb effects are small enough to be neglected, close to 1.5 for the largest nuclei.

Note that if a large nucleus decays by emitting an α, the nucleus left behind has an even greater ratio of neutrons to protons, since $\alpha = 2p+2n$. This implies that it might be able to go to a lower energy state if it could trade a neutron for a proton. In fact, this can sometimes be done. The neutron itself is not a stable particle. A neutron alone in space will only last about ten minutes before ejecting an electron, leaving a proton behind. This is energetically possible, because the mass of the neutron exceeds the sum of the masses of the proton and electron.

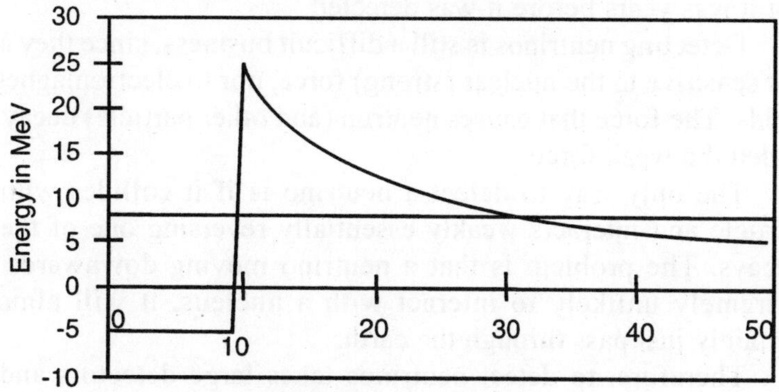

Fig. Distance in Fermis

The process is called β-decay. It can also occur in a nucleus, but now the different binding energies of the initial neutron and the final proton must be factored in to decide if it will go. Other things being equal, the more neutrons there are present in the nucleus, the more likely it is to β-decay. Actually, β-decay was a big puzzle when it was first carefully observed, because it was discovered that identical nuclei β-decaying to the same final states ejected electrons over a range of energies, up to a maximum corresponding to the energy difference between the initial and final nuclei. This led to the suggestion that maybe, after all, energy wasn't conserved in every process in nature.

But physicists are a conservative lot when it comes to energy conservation, and Pauli suggested that maybe $n \rightarrow p+e$ wasn't the whole story some other particle was being created at the same time, carrying away the balance of the energy, but *undetected*. This meant it had no charge, no mass, and, unlike the photon, no electric interaction. Fermi called it the *neutrino*, the little neutral one. Another reason the neutrino seemed necessary was that the initial neutron had spin one-half, as did the final proton *and* the emitted electron. It was difficult to see how angular momentum could be conserved if no other particle were emitted, because even if the final proton and electron had some orbital angular momentum, that comes in whole-number amounts, so how could two half-integer spins and an integer angular momentum add to be equal to the initial half-integer spin? It turned out they couldn't, the neutrino really was there all along, it had spin one-half itself, but it was years before it was detected.

Detecting neutrinos is still a difficult business, since they are not sensitive to the nuclear (strong) force, nor to electromagnetic fields. The force that causes neutron (and other particle) decay is called the weak force.

The only way to detect a neutrino is if it collides with a particle and interacts weakly essentially reversing one of these decays. The problem is that a neutrino moving downwards is extremely unlikely to interact with a nucleus,. it will almost certainly just pass through the earth.

Therefore, to detect neutrinos takes large detectors and a strong flux of neutrinos. Much effort has been expended detecting neutrinos from the sun, since they are our only window into the nuclear processes taking place deep in the sun—they can get out unscathed, unlike photons or any other particle emitted. In fact, it is found that the number detected is less by a factor of two or three from that predicted using standard analyses of the solar nuclear processes.

It turns out that in fact the neutrino is not quite massless, and there are actually three kinds of neutrinos! Furthermore, as a neutrino passes through matter, there is a nonzero probability of it changing from one kind to another. The solar nuclear processes generate what are called electron neutrinos, and the initial efforts at neutrino detection looked for that kind.

More recent detectors count the other two kinds (mu neutrinos and tau neutrinos) as well, and the total number of neutrinos detected agrees well with the prediction of the total number generated in the sun, so it is now believed that some of the original electron neutrinos change types traveling from their point of origin to the sun's surface.

Anyway, back to the radioactive decay of U^{238}. It emits an α to become Th^{234}, then β decays twice in succession to become

U^{234}. This is now relatively proton-rich, and in fact a decays five times in succession to become a form of lead,

$$U^{234} \rightarrow Th^{230} \rightarrow Ra^{226} \rightarrow Rn^{222} \rightarrow Po^{218} \rightarrow Pb^{214}$$

Pb^{214} undergoes two β decays, an α decay and two more β decays before it reaches stability at Pb^{206}. The sequence above is through metallic elements with one exception—Rn, radon, is a gas because, atomically speaking, it has completely filled electron shells, and so is chemically unreactive, and gaseous at ordinary temperatures. It emits its α particle in a period of days, which gives it time to diffuse out of the rock into somebody's basement.

γ-DECAY

The three forms of radiation emitted by decaying nuclei, α, β and γ, were named before the particles being emitted were fully identified. It was only realized later that the β-rays were electrons. The γ rays are just high-energy photons, of order 100 keV to a few MeV.

Emission of γ rays is similar (but at much higher energies) to emission of photons by excited states of atoms. The nucleus can be excited by having just emitted an α or β, or by colliding with another nucleus, or being bombarded by neutrons, say. All these events can lead to a nucleus in which the charge distribution is oscillating, and electromagnetic radiation ensues.

THE DROPLET MODEL

In some ways, the binding of nucleons in a nucleus is reminiscent of that of molecules in a drop of water—the force is attractive between near neighbors. Notice how different this is from the electrons in an atom, held in place by a large central attracting object. This "droplet" model of the nucleus is useful for understanding nuclear fission. If the droplet is somehow caused to vibrate vigorously, it might break into two, or more, smaller nuclei.

For the U^{235} nucleus (only 0.7% in naturally occurring uranium), if a slow neutron gets too close, the attractive nuclear force pulls it towards the nucleus so strongly that in the resulting collision, the nucleus breaks into two smaller ones, and two or three neutrons are ejected (since the smaller nuclei formed have lower proportions of neutrons). There is also substantial energy release—about 200 MeV. The much more common U^{238} nucleus also attracts free neutrons, but not so strongly, and it survives neutron capture without fission.

A neutron has to be moving very slowly to be captured by U^{235}. On the other hand, when U^{235} fissions the emitted neutrons are moving very fast. To get a *chain reaction* in a nuclear reactor, the main problem is to slow down the emitted neutrons before they leave the reactor.

If they can be slowed down successfully, neutrons emitted by one fission will initiate further fissions, giving continuous—perhaps increasing—energy output. However, the neutrons bounce off the heavy uranium nuclei with almost no energy loss—to get a particle to lose energy in a collision, it must collide with something approximately its own size.

Water is often used, the neutrons lose energy bouncing off the protons in the hydrogen atoms. Unfortunately, the neutrons often stick to the protons, forming deuterons, so to keep the reaction going the uranium must be enriched—the U^{235} content boosted to one to four percent. Of course, care must be taken to ensure that the chain reaction isn't *too* successful moderators, which absorb neutrons, are put into the reactor if the rate of fission increases beyond the planned level. Needless to say, moderators are not used in military applications of this concept.

Index